Multi-Component Force
Sensing Systems

Series in Sensors
Series Editors: Barry Jones and Haiying Huang

For more information about this series, please visit: https://www.routledge.com/Series-in-Sensors/book-series/TFSENSORS

Multi-Component Force Sensing Systems

Qiaokang Liang

CRC Press
Taylor & Francis Group
Boca Raton London New York

CRC Press is an imprint of the
Taylor & Francis Group, an **informa** business

First edition published 2021
by CRC Press
6000 Broken Sound Parkway NW, Suite 300, Boca Raton, FL 33487-2742
and by CRC Press
2 Park Square, Milton Park, Abingdon, Oxon, OX14 4RN

ISBN: 978-0-367-50240-9 (hbk)
ISBN: 978-1-003-05253-1 (ebk)

Typeset in Minion
by KnowledgeWorks Global Ltd.

Contents

CHAPTER 1

Introduction

Autonomous and intelligent control is one of the main trends in research and development of robotics. The perception and feedback are necessary means for robot intelligent behaviors, and the level of intelligent control depends largely on the comprehensive understanding of the environment in which the perception and interaction system is located. High-quality visual and force perception information is the most important condition for the system to complete the operation. Although the vast majority of informational acquisition comes from the visual system, human acquisition and understanding of force and haptic information occupy 2/3 of the brain resources. Therefore, the force perception and interaction system is one of the most complex and comprehensive perception methods. How to efficiently and reliably acquire and understand the force interaction between the robot and the working environment is an urgent need for intelligent robots to achieve reasonable human-machine interaction and intelligent control. For example, when a collaborative robot equipped with a force perception system completes human-machine collaborative tasks such as zero-force teaching and flexible control, the control system of the collaborative robot will adjust its state in time according to the interaction in order to take into account the position servo stiffness of the multi-degree-of-freedom robot control system and the flexibility of the interaction between the collaborative robot and the collaborators. Therefore, the achievement of active compliance control relies on the force perception, i.e., the force/torque feedback information of the multi-component force sensing system, with which the robotic system can complete the traction follow-up, space curve/surface tracking, constant force operation, flexible assembly, and other efficient motion control strategies.

1.1 WHAT IS A MULTI-COMPONENT FORCE SENSING SYSTEM?

A force sensing system can obtain the interaction force between the robot and the external environment during operation. It is one of the most important perceptions of intelligent robots. Force sensing systems with multi-component perception capability can simultaneously sense the force or moment information in two or more directions in the Cartesian coordinate system, and then realize the information perception such as force, touch, tactile, and slip of the manipulations. Specifically, a six-component force can monitor six components of force terms along the x-, y-, and z-axis (Fx, Fy, and Fz) and the torque terms about x-, y-, and z-axis (Mx, My, and Mz) simultaneously. Force and torque information is always adopted as feedback to obtain a closed-loop control system. As shown in Fig 1.1, the manipulation force and torque between the robot and environment are detected by the multi-component force sensing system, which is used to process the force and position control, i.e., a strategy of robot compliance control.

The measurement chain of multi-component force sensing systems consists of several elements, regardless of the adopted principles. Compared to force sensors, the multi-component force sensing systems are characterized by:

- Multi-component force transducer consists of an elastic structure and corresponding measuring elements. The applied load (force and/ or torque along an arbitrary axis) acts on the elastic structure, and the measuring element and circuit will transform corresponding physic

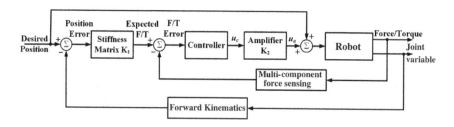

FIGURE 1.1 Multi-Component F/T Sensing Systems in Force Servo Control Strategy for Robot.

variations (such as wavelength shift, displacement, strain, potential, etc.) occurred on the elastic structure into the electrical quantity.

- The obtained electrical quantities such as changes in voltage and current are usually small; therefore, instrumentations with amplification function are adopted, which are always placed close to the measuring circuit.

- Data acquisition devices with functions such as analog-to-digital conversion, signal conditioning (filtering, isolation, temperature compensation, linearization, creep correction, etc.) are equipped as the interface between the multi-component force transducer and a computer or an embedded microcomputer system.

- The computer or microcomputer system is adopted to processing (calculation, calibration, and software filtering), storing, and visualizing the measured force information.

1.2 CLASSIFICATION OF MULTI-COMPONENT FORCE SENSING SYSTEMS

The earliest detection methods used in force sensing systems are capacitive, resistive strain, and piezoelectric. With the continuous development of detection and sensing technology, more and more detection methods such as fiber grating, inductive, pneumatic, and electromagnetic types are used in force sensing systems. The advantages and disadvantages of various methods (as shown in Table 1.1) determine their applications in different situations.

Among various principles for multi-component force sensing systems, the most popular approach relies on resistive strain due to its characters such as theoretical maturity, simple processing circuit, adjustable resolution, and high reliability. Another popular approach to measure multi-component is the capacitive method, which shows high resolution, excellent durability, and wide frequency bandwidth. Piezoelectric force sensing systems with piezoelectric effect have received considerable attention due to its advantages such as wide detection range, superior stiffness, high-frequency response, and accuracy. Nowadays, more and more force sensing systems based on metal foil strain gauge are replaced with piezoresistive-based approaches, which feature with increased gauge factor and higher sensitivity. State-of-the-art force sensing systems adopt

TABLE 1.1   Disadvantages and Advantages of Different Force Sensing Method

Sensing Principle	Transduction Effect	Advantages	Disadvantages
Magnetoelectric	hall effect	wide dynamic range; low power consumption; high reliability	poor interchangeability; large nonlinear error; low resolution
Capacitive	capacitance variation due to a load	high sensitivity and resolution; large bandwidth; robustness; drift-free; durability;	complex electronic circuits; stray capacitance; edge effect
Resistive	conductivity variation due to a load	theoretical maturity; adjustable resolution; high reliability; maintenance-free	higher power consumption; rigid and fragile; scarce reproducibility
Inductive	magnetic coupling variation due to a load	high sensitivity and resolution; linear output;	lower-frequency response; poor reliability
Optoelectronic	refractive index variation due to a load	good reliability; wide detection range; noncontact	non-conformable; hard to construct dense arrays
Piezoelectric	piezoelectric effect	High-frequency response; higher accuracy; high sensitivity and dynamic range; high stiffness	charge leakages; poor spatial resolution; deteriorations of voltages or drifts in the presence of static forces

Fiber Bragg Grating (FBG) or other optoelectronic approaches to detect force and torque with good reliability, wide detection range, and Magnetic Resonance Imaging (MRI) compatibility.

1.3 STATE OF THE ART AND TRENDS IN MULTI-COMPONENT FORCE SENSING SYSTEMS

As shown in Table 1.2, the research and development of multi-component force sensing systems have been experiencing steady and rapid growth during the past decades. Since 1980, force sensing systems have played an important role in the development and application of intelligent manufacturing and intelligent equipment such as robot-assisted surgical robots, flexible assembly systems, automatic manipulator, cutting force detection, medical-aided rehabilitation training system, etc.

TABLE 1.2   Count of Papers Related Multi-Component Force Sensing Systems from 1980

Year	Ieee Library	Compendex	Asme Digital Collection	Spie Digital Library	Springer-Link
1980-1989	2	0	0	0	-
1990-1999	126	61	2	6	110
2000-2009	608	649	171	173	1144
2010-present	1576	1023	769	547	1937

Table 1.3 summarizes some collections of contributions to the state of the art of research related to multi-component force sensing systems.

Dai et al. [1] proposed an integrated tri-axial force sensor system for robotic minimally invasive surgery based on the capacitive approach, which was demonstrated for the first time a multi-component force sensing system can be integrated into commercial surgical graspers. The proposed system was supposed to demonstrate that real-time multi-component force feedback can improve the speed or outcomes of gastrointestinal surgical procedures. Dwivedi et al. [2] designed a soft magnetic three-component force sensor for a force-controlled pick-and-place experiment. A novel pyramid-shaped tactile unit based on a tri-axis hall element and a magnet was proposed, and its non-linear output was characterized by a neural network. Experimental results show that the multi-component force information can be used to achieve a force controlled manipulation with fragile objects. A. Hamed et al. [9] designed a bio-inspired three-component force sensor based on a three-DOF decoupled parallel mechanism for human-robot interaction purposes. With the proposed force sensor, fixed admittance control and active admittance control were applied for the human-robot interaction. T. Takeshita et al [16] developed a two-component optical force sensor with a size of $6 \times 6 \times 8$ mm. The proposed sensor with an elastic frame and optical sensors chip was capable of detecting shearing force along two axes independently with a resolution of 0.063 N. Hunan university, in collaboration with other academic institutions such as the Institute of Intelligent Machines, Chinese Academy of Sciences, and York University has been an active player in the development of multi-component force sensing systems. Qiaokang Liang et al. [20] have designed and developed a series of multi-component force sensing systems, such as three- [17], four- [18], five- [19], and six-component sensing systems for flexible bio-robotic foot/ankle, underwater robot manipulator, coordinate measuring machine (CMM), compliant parallel mechanism (CPM), respectively.

TABLE 1.3   State of the Art in Multi-Component Force Sensing System

Researcher	Method	Calibration	Dimension(Mm) & Component	Sensitivity & Measurement Range	Principle
2017 & Dai Y. et al. [1]	MEMS	dynamometer	3 × 6 mm & 3	0.6218fF/N & 4 N	capacitive
2018 & A. Dwivedi et al. [2]	silicone rubber molding	force sensor TAL220	12 × 12 × 8 & 3	11.7 mN & ±1.5 N	magnetoelectric
2017 & T. Nagatoma et al. [3]	liquid alloy casting	pressure testing machine	10 & 3	0.49% & 0~5 N	capacitive
2017 & H. Choi et al. [4]	silicone rubber molding	force sensor USL08-H6	$\Phi 40 \times 7$ & 3	0.943%F.S & 0~13 N	pneumatic chambers
2017 & V. Grosu et al. [5]	machining	dynamometer Vishay1550	80 × 96 & 3	1.6 mV/N(Fz)/& 300 N	resistive
2016 & Kim et al. [6]	PCB	Nano17, ATI	<10 mm & 3	0.15 N & 8 N	capacitive
2018 & Xiong et al. [7]	Machining	weight	$\Phi 72 \times 26$ & 3	11.259 PM/N & ±60 N	FBG
2019 & Wolterink et al. [8]	3D print	scales	22 × 12 & 3	100Ω/N & 10 N	resistive
2020 & Hamed et al. [9]	MEMS	force sensor	12.6 × 11.4 & 3	0.01 N & 4 N	resistive
2013 & Kan et al. [10]	silicon micromachining	uniaxial piezoresistive reference cantilever	2.54 × 1.76 × 0.015 & 3	1.5 N m^{-1}, 0.1 μN & n. a.	piezoresistive
2012 & Cullinan et al. [11]	micro-fabrication and self-assembly	micrometer with resolution of 1 μm	2.5 × 0.035 × 0.01 & 3	0.79 mV μN^{-1} & 100 μN	CNT piezoresistive
2011 & Cappelleri et al. [12]	laser etching and wet etching	calibrated AFM cantilever and FEA	3 × 1.63 × 0.075 & 2	26 pixels/ μN & 0-50 μN	vision-based
2012 & Brookhuis et al. [13]	bulk silicon micromachining and silicon fusion bonding	calibrated reference sensor	9 × 9 × 1(PCB chip) & 3	16 PF/N in F_z and 2.7PF/N mm in M_x and M_y & 50N in F_z, ±25 N mm in M_x and M_y	capacitive
2012 & Estevez and Bank [14][15]	bulk silicon micromachining	a commercial mono-axial force probe	3 × 1.5 × 0.03 & 6	100 μN & 4-30 mN and 4-50 μN m in forces and moments respectively	piezoresistive

J. Park et al. [21] presented a fingertip force sensor based on capacitive approach for glove-type assistive devices. The proposed sensor was capable of detecting the grip force with an accuracy of 3.75% after calibration, regardless of various posture conditions.

1.4 CHALLENGES AND FUTURE DIRECTIONS OF MULTI-COMPONENT FORCE SENSING SYSTEMS

Recently, multi-component force sensing systems have been attracting an increased level of attention, and continue efforts have been made on the design, investigation, and fabrication of task centered sensing systems by employing different transduction techniques.

1.4.1 Elastic Structures Based on Sophisticated Structures

One of the most critical components of multi-component force sensing systems is the elastic structure, which acts as the interface between the applied load and the measuring element of the measurement circuit. Monolithic structures such as cross beams, membranes, thin-walled hollow column, compliant parallel mechanism, pillar, ring, etc. are popular for multi-component force sensing systems, in view of the structural complexity and simplicity of theoretical modeling. An appropriate design of elastic structures is always characterized by high sensitivity, measurement isotropy, low coupling output, simple structure, low error (such as nonlinearity, hysteresis, and repeatability errors), and minimum dimension (especially small height).

Recently, compliant parallel mechanisms are considered as an excellent candidate for the elastic structure for a multi-component force sensing system due to their characters:

- Theoretical and experimental maturity. The analysis and experiment on compliant parallel mechanisms (such as dynamics, stiffness analysis, kinematics character, etc.) have been investigated regularly by researchers and have arrived at a level of maturity both theoretically and experimentally.

- Isotropy of measurement among components. The global stiffness matrix of compliant parallel mechanisms expresses the relationship between the infinitesimal movement of the end effector and the applied load. Measurement isotropy among components can be achieved through static analysis.

- Low coupling error compared to traditional elastic structures. Compliant parallel mechanisms detect the six force/torque components with several limbs with more active sensing portions, which therefore enable the possibility of low coupling output with high sensitivity and stiffness in the meantime.

Practical structures and geometry dimensions of multi-component force sensing systems generally determined the required measurement range, component number, maximum allowable error, response time, etc. A number of design approaches have proposed about the high-performance multi-component force sensing systems. However, the majority of them rely on the experience of designers or direct optimization algorithms. Therefore, the efficient design and optimization approach for multi-component is still a challenge in the engineering and scientific communities. Unlike the sophisticated structures, modular structures (such as cantilever beams and circular diaphragms) with several active sensing portions are also adopted by researchers.

1.4.2 Automatic Calibration and Decoupling

Calibration experiments are always used to determine the output values of the multi-component force sensing system that corresponds to the applied load. By individually applying known reference load (from minimum to maximum then decrease to the minimum with a constant increment) onto the force sensing system and monitoring the corresponding system output voltage with a cyclical pattern, the relationship between the applied load and the corresponding output can be finally determined. Generally, almost all the output of multi-component force sensing system are confronted by high coupled interference errors among components due to their monolithic structural element, manufacturing errors, and signal processing algorithms, i.e., the unexpected output of other components along with the correct output occurs due to the applied force or torque component along a single axis. Decoupling refers to eliminating or decreasing the coupling error among components of multi-component force sensing systems. Specifically, decoupling methods can be divided into two categories, namely, those that realize elimination coupling error with circuit hardware and those with decupling algorithms. The study on decoupling algorithms mainly includes static decoupling (including linear and nonlinear decoupling) and dynamic decoupling (also referred to as dynamic decoupling and compensation). In this book, only the decoupling based on linear and nonlinear algorithms will be considered.

1.4.3 Integrated Micro Multi-Component Force Sensing Systems

A number of micro sensors and other MEMS technologies are being embedded into multi-component force sensing systems to act as integrated systems, which may consist of a micro elastic structure, a device package that encases the system into a complete device, and an integrated measuring circuit for signal conditioner as well analog/digital signal processing. One of the difficulties in integrated micro multi-component force sensing systems is their packaging and special interfaces because the elastic structure always needs a compliant space to deform and then detects the corresponding strain. Additionally, how to integrate low-noise amplification, high-performance analog-to-digital conversion, and compensation component into the system is yet difficult and challenging.

1.4.4 Full Component Overload Protection and Stiffness

Overload and stiffness have been major problems with previous multi-component force sensing systems. The overload protection prevents damage to the elastic structure when an excess load is placed onto the systems. Almost all force sensing systems measure force and torque without overload protection or only with several component protections. No mature overload protection for full component force and torque of the force sensing systems is available until now. The stiffness of automatic force control systems depends on the weakest component with minimum stiffness in the force-flux flow, which is always the force sensing system. Therefore, modern control systems can benefit from force sensing systems with higher stiffness. While higher stiffness always accompanies with lower sensitivity and decreased resolution of the sensing systems. To solve these problems, scientists and engineers use simulation-driven design, multi-objective optimization, and optimization exploration to design and develop stiffer force sensing systems with all-component overload protection.

1.4.5 Dynamic Behavior Characters

Multi-component force sensing systems in automatic control systems actually behave dynamically when they are applied with a load. The dynamic behavior characters refer to measures of the ability of sensing system to track rapid changes in the applied load. Specifically, the force sensing systems are only capable of detect force and torque within a limited frequency range. Additionally, amplitude and phase distortion characteristics are critical dynamic parameters of the multi-component force sensing systems. Furthermore, force sensing systems always need a period of time to follow the occurred input load, and this response time is usually stated as the time that the system reaches a new settled value from its previous status. All traditional multi-component force sensing systems always have poor dynamic response performance, and the average response time of most existing multi-component force sensing systems is around 5–20 ms, which is required further improvement. Moreover, calibration, decoupling, and compensation of multi-component force sensing systems in a dynamic manner should be considered in the detailed design period.

FBG and piezoelectric force sensing systems have demonstrated remarkable benefits in dynamic detection of strain, pressure, temperature, force, torque, vibration, deformation, and bending. Therefore, future force sensing systems with excellent dynamic behavior characters may emerge with FBG or piezoelectric approaches.

REFERENCES

1. Dai Y, Abiri A, Liu S, Payder O, et al. Grasper integrated tri-axial force sensor system for robotic minimally invasive surgery[C]//2017 39th Annual International Conference of the IEEE Engineering in Medicine and Biology Society (EMBC). IEEE, 2017: 3936–3939.
2. Dwivedi A, Ramakrishnan A, Reddy A, Patel K, et al. "Design, Modeling, and Validation of a Soft Magnetic 3-D Force Sensor[J]." IEEE Sensors Journal 18.9 (2018): 3852–3863.
3. Nagatomo T, Miki N. "Three-Axis Capacitive Force Sensor with Liquid Metal Electrodes for Endoscopic Palpation[J]." Micro & Nano Letters 12.8 (2017): 564–568.
4. Choi H, Jung P G, Jung K, Kong K. Design and fabrication of a soft three-axis force sensor based on radially symmetric pneumatic chambers[C]//2017 IEEE International Conference on Robotics and Automation (ICRA). IEEE, 2017: 5519–5524.
5. Grosu V, Grosu S, Vanderborght B, Lefeber D, Rodriguez-Guerrero C.. "Multi-Axis Force Sensor for Human–Robot Interaction Sensing in a Rehabilitation Robotic Device[J]." Sensors 17.6 (2017): 1294.
6. Kim U, Kim Y B, Seok D Y, J So, Choi H R. Development of surgical forceps integrated with a multi-axial force sensor for minimally invasive robotic surgery[C]//2016 IEEE/RSJ International Conference on Intelligent Robots and Systems (IROS). IEEE, 2016: 3684–3689.
7. Xiong L, Jiang G, Guo Y, Liu H. "A Three-Dimensional Fiber Bragg Grating Force Sensor for Robot[J]." IEEE Sensors Journal 18.9 (2018): 3632–3639.
8. Wolterink G, Sanders R, Krijnen G. A flexible, three material, 3D-printed, shear force sensor for use on finger tips[C]//2019 IEEE SENSORS. IEEE, 2019: 1–4.

9. Hamed A, Masouleh M T, Kalhor A. Design & Characterization of a Bio-Inspired 3-DOF Tactile/Force Sensor and Implementation on a 3-DOF Decoupled Parallel Mechanism for Human-Robot Interaction Purposes[J]. Mechatronics 2020, 66: 102325.

10. Kan, T, Takahashi H, Binh-Khiem N, Aoyama Y, et al., "Design of a Piezoresistive Triaxial Force Sensor Probe Using the Sidewall Doping Method," Journal of Micromechanics and Microengineering, 23.3, pp. 1–7, 2013.

11. Cullinan M. A., Panas R. M.and Culpepper M. L. "Design and Fabrication of a Multi-Axis MEMS Force Sensor With Integrated Carbon Nanotube Based Piezoresistors," Nanotechnology, 2, pp. 302–305, 2011.

12. Cappelleri D. J., Krishnan G, Kim C, Kumar Vand Kota S. "Toward the Design of a Decoupled, Two-Dimensional, Vision-Based μN Force Sensor," ASME J. Mech. Robot., 2, pp. 1–9, May 2010.

13. Brookhuis R. A., Lammerink T. S. J., Wiegerink R. J., de Boer M. J. and Elwenspoek M. C. "3D Force Sensor for Biomechanical Applications," Sensor Actuator Phys., 182, pp. 28–33, 2012.

14. Estevez P., Bank J., Porta M., Wei J., Sarro P. M., Tichem M. and Staufer U. "6 DOF Force and Torque Sensor for Micro-Manipulation Applications," Procedia Engineering, 25, pp. 39–42, 2011.

15. Estevez P., Bank J. M., Porta M., Wei J., Sarro P. M., Tichem M. and Staufer U. "6 DOF Force and Torque Sensor for Micro-Manipulation Applications," Sensor Actuator Phys., 186, pp. 86–93, 2012.

16. Takeshita T., Harisaki K., Ando H., Higurashi E., Nogami H. and, Sawada R. (2016). Development and Evaluation of a Two-Axial Shearing Force Sensor Consisting of an Optical Sensor Chip and Elastic Gum Frame. Precision Engineering, 45, 136–142.

17. Liang Q., Zhang D., Song Q. and Ge Y. (2010, August). Design and evaluation of a novel flexible bio-robotic foot/ankle based on parallel kinematic mechanism. In 2010 IEEE International Conference on Mechatronics and Automation (pp. 1548–1552). IEEE.

18. Liang Q., Zhang D., Song Q., and Ge Y. "A Potential 4-D Fingertip Force Sensor for an Underwater Robot Manipulator." IEEE Journal of Oceanic Engineering 35.3 (2010): 574–583.

19. Liang Q., Zhang D., Wang Y., and Ge Y., et al. "Development of a Touch Probe Based on Five-Dimensional Force/Torque Transducer for Coordinate Measuring Machine (CMM)." Robotics and Computer-Integrated Manufacturing 28.2 (2012): 238–244.

20. Liang Q., Zhang D., Chi Z., Song Q., et al., "Six-DOF Micro-Manipulator Based on Compliant Parallel Mechanism With Integrated Force Sensor." Robotics and Computer-Integrated Manufacturing 27.1 (2011): 124–134.

21. Park J., Heo P., Kim J., Na Y., et al. "A Finger Grip Force Sensor With an Open-Pad Structure for Glove-Type Assistive Devices." Sensors 20.1 (2020): 4.

CHAPTER 2

Multi-Component Force Sensing Systems Based on Different Detection Techniques

The most commonly studied and widely used detection methods and techniques for multi-component force sensing systems are derived from the development of various detection principles, such as resistive, capacitive, piezoelectric, inductive, magnetic, and optical approaches.

2.1 MULTI-COMPONENT RESISTIVE FORCE SENSING SYSTEMS

The most widely adopted multi-component force sensing systems rely on the resistive measurement technique because electrical resistance is one of the easiest properties to be measured with a moderate cost. Multi-component resistive force sensing systems have many excellent characteristics, such as simple construction, high reliability, adjustable resolution, and cost effectiveness which have made them the ideal option for design and application of multi-component force sensing systems [1]. The following analysis and approach to design and development multi-component force sensing is based on our previous work [2].

According to the piezo-resistive behavior, the relative resistance change of a piezo-resistor is given by [3]:

$$\frac{\Delta R}{R} = \xi'_{11}\sigma_1 + \xi'_{12}\sigma_2 + \xi'_{13}\sigma_3 + \xi'_{14}\vartheta_4 + \xi'_{15}\vartheta_5 + \xi'_{16}\vartheta_6 \tag{1}$$

where the primed quantities ξ'_{1i} (i=1, 2, 3, 4, 5, 6) represent the piezo-resistive coefficients referred to arbitrarily oriented axes, σ_j (j=1, 2, 3) and ϑ_k (k=4, 5, 6) denote the normal stresses and shear stresses with respect to the applied load, respectively.

In addition, a strain gauge based sensing element is attached onto the elastic structure by a suitable adhesive, and its resistance varies in accordance with the applied force or moment. The resistance change rate can be written as

$$\frac{\Delta R}{R} = G\varepsilon \tag{2}$$

where G is the gauge factor of the strain gauge, and ε represents the strain caused by the applied force or moment.

Wheatstone Bridge configuration circuits are usually employed as measurement circuit by electrically connecting the strain gauges in order to provide excellent linearity and improved sensitivity. The output voltage of the bridge can be expressed as

$$\Delta U = \frac{1}{4}V_e G(\varepsilon_1 - \varepsilon_2 + \varepsilon_3 - \varepsilon_4) \tag{3}$$

where V_e denotes the voltage excitation source, and ε_i represents the strain in the gauge of the ith leg of the full-bridge circuit.

As shown in Fig 2.1, three groups of strain gauges are bonded onto the circular membrane. Specifically, two stain gauges (R_{x1} and R_{x2}) are mounted at the spots with the maximum strain on the membrane along the x-axis to detect F_x (force component along the x-axis). Two strain gauges (R_{y1} and R_{y2}) are mounted at the spots with the maximum strain on the membrane along the y-axis to detect F_y (force component along the y-axis). Two stain gauges (R_{z1} and R_{z2}) are mounted at the spots with the maximum strain on the membrane along the 45 degrees to the x-axis to detect F_z (force component along the z-axis). Detection with train

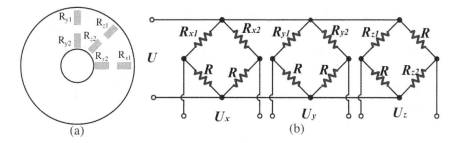

FIGURE 2.1   Multi-Component F/T Sensing Systems Based on Resistive Technique.

gauges always involves quantities smaller than a few milli-stain, therefore the corresponding changes in their electric resistances are tiny. Therefore, the Wheatstone bridge circuits are always adopted to electrically connect the strain gauges with strains bonded on non-stress regions (R), as shown in Fig 2.1 (b).

Assuming that all strain gauges have the same initial resistance, the output of the circuit can be expressed as the following result.

$$U_x = \frac{U}{4}\left(\frac{\Delta R_{x1}}{R_{x1}} - \frac{\Delta R_{x2}}{R_{x2}} + \frac{\Delta R}{R} - \frac{\Delta R}{R} \right) = \frac{U}{2}\left(\frac{\Delta R_{x1}}{R_{x1}} \right) \tag{4}$$

$$U_y = \frac{U}{4}\left(\frac{\Delta R_{y1}}{R_{y1}} - \frac{\Delta R_{y2}}{R_{y2}} + \frac{\Delta R}{R} - \frac{\Delta R}{R} \right) = \frac{U}{2}\left(\frac{\Delta R_{y1}}{R_{y1}} \right) \tag{5}$$

$$U_z = \frac{U}{4}\left(\frac{\Delta R_{z1}}{R_{z1}} - \frac{\Delta R}{R} + \frac{\Delta R_{z2}}{R_{z2}} - \frac{\Delta R}{R} \right) = \frac{U}{2}\left(\frac{\Delta R_{z1}}{R_{z1}} \right) \tag{6}$$

Multi-component resistive force sensing systems are commonly used in many applications. However, systems may become rigid and fragile due to their mechanical structure and bonding process of gauges. Additionally, silicon strain gauges also have problems such as temperature sensitivity and drift.

When the forces to be measured Fx, Fy or Fz act on the sensing system, the circular loop connected with the upper adapter floats, while the base frame connected with the lower adapter acts as fixed support. Consequently, corresponding elastic deformation and strain will take place on the lower diaphragm.

In general, a multi-component force sensing system generally experiences with a critical drawback in the form of tradeoffs. These tradeoffs are typical in mechanical design and consist of a compromise between different performances, such as sensitivity and stiffness, isotropy and coupling, static and dynamic performances. For example, optimum absolute sensing system sensitivity corresponds to the desire of generating as much elastic strain as possible in the elastic element. However, both the sensing system sensitivity and general stiffness strongly depend on the geometrical parameters of the elastic element structure. Specifically, maximizing the sensing system sensitivity is in contradiction with maximizing the general stiffness. Consequently, the optimal design and development of a multi-component force sensing is complicated and cannot be realized through direct optimization methods.

The Simulation Driven Optimization (SDO) enables the product design and development process with more efficient. Design Exploration (DE) can simulate the response pattern with input parameter variation. The optimal choice of the input parameters such as dimensions of the elastic element can be then decided.

Input parameters are set as the geometric dimensions of the diaphragm of the elastic element. Additionally, the maximum deformation, the minimum strain, the maximum stress, and the primary response of the elastic element are set as the output of the optimization process. Table 2.1 illustrated that the design parameter ranges and the objectives for output variables, and all targets are treated with identical important

TABLE 2.1 Design Parameters and Output Variables

	Design Parameters			**Output Variables**		
Parameter	Extended Circle Diameter	Diaphragm Thickness	Maximum Total Deformation (mm)	Minimum Elastic Strain (10-6mm/ mm)	Maximum Equivalent Stress (MPa)	The Primary Response Frequency (Hz)
VARIABLE	P1	P2	P5	P6	P7	P8
Bound/ Objective	40<P1<50	1<P2<4	P5<0.03	P6>100	P7<60	P8>500
Candidate 1	49.55	1.055	0.0273	159.884	39.572	744.3
Candidate 2	46.35	1.031	0.0265	134.272	54.719	1210.4
Candidate 3	47.95	1.149	0.0192	130.072	40.916	1043.7

TABLE 2.2 Variations of Resistances of Strain Gauges

		Fx	Fy	Fz	Mx	My	Mz
F_x-bridge	R_1	+	0	+	0	0	+
	R_2	-	0	-	0	0	+
	R_3	+	0	-	0	0	+
	R_4	-	0	+	0	0	+
F_y-bridge	R_5	0	+	+	0	0	+
	R_6	0	-	-	0	0	+
	R_7	0	+	-	0	0	+
	R_8	0	-	+	0	0	+
F_z-bridge	R_9	+	+	+	0	0	+
	R_{10}	-	-	-	0	0	+
	R_{11}	+	+	-	0	0	+
	R_{12}	-	-	+	0	0	+

weight. Specifically, the first as well as the third targets about the stress and deformation are used to make sure the elastic element works in elastic region of the material with accuracy, linearity, and satisfying stiffness. While the second and fourth targets about the strain and dynamic response behaviors can enable the force sensing system with high resolution, excellent sensitivity, and fine dynamic performance. Dozens of design candidates are obtained by the method, and the most appropriate ones are listed in Table 2.1. It is notable that all these targets can be fulfilled with these candidates, in which the candidate set one is adopted in our design.

When the load is applied onto the force sensing system, the electric resistance of strain gauges bonded on the elastic element will change. Table 2.2 illustrates the electrical resistance variations according to the applied six components of the load. Specifically, the symbols "+" and "−" represent an increase and decrease of the electrical resistance, respectively. While the symbols "=" and "0" represent identical and no variation, respectively. It is revealed that the FEA results are identical to that obtained by the theoretical model.

Calibration and decoupling experiments of the proposed force sensing system are implemented. Table 2.3 illustrated that the outputs of the force sensing system under different loads. Additionally, the force sensing system output exhibits an excellent symmetry about the origin of the coordinate system.

TABLE 2.3 The Outputs of the Sensor

Component	Fx	Fy	Fz
Applied load	-	-	-
Corresponding output	-	-	-
Applied load	−200 N	−200 N	0 N
Corresponding output	−4.48 V	−4.35 V	−.0034 V
Applied load	−150 N	−150 N	50 N
Corresponding output	−3.41 V	−3.31 V	0.73 V
Applied load	−100 N	−100 N	100 N
Corresponding output	−2.35 V	−2.28 V	1.40 V
Applied load	−50 N	−50 N	150 N
Corresponding output	−1.27 V	−1.25 V	2.07 V
Applied load	0 N	0 N	200 N
Corresponding output	−.00153 V	−.00168 V	2.72 V
Applied load	50 N	50 N	250 N
Corresponding output	1.28 V	1.26 V	3.35 V
Applied load	100 N	100 N	-
Corresponding output	2.40 V	2.28 V	-
Applied load	150 N	150 N	-
Corresponding output	3.43 V	3.32 V	-
Applied load	200 N	200 N	-
Corresponding output	4.5 V	4.37 V	-
Applied load	-	-	-
Corresponding output	-	-	-

The performance indexes of the designed force sensing system based on the resistive approach are listed in Table 2.4. In comparison with the existing force sensing systems, the proposed force sensing system shows superiorities in the coupling and nonlinearity errors. However, from the experimental results, the proposed sensing system still suffers from a slight coupling effect due to the monolithic structure of the elastic element and the bonding error of the strain gauges.

TABLE 2.4 The Performance of the System

Component	Fx	Fy	Fz
Sensitivity	.02245 VN^{-1}	.02180 VN^{-1}	0.0134 VN^{-1}
Maximum coupling error	1.07%	1.38%	0.41%
Maximum nonlinearity error	1.75%	1.94%	1.87%

A detailed discussion of the state-of-the-art multicomponent force sensing systems may be found in Ref. [1]. Different types of resistive force sensing systems have been explored. D. Singh et al. [3] proposed a soft 3D printed resistive force sensor for rehabilitation robotics with a measurement range of 0 - 13 N. The proposed soft force sensor was composed of a sensor base and a deformable head. Electrical resistance will change when the soft force sensor contact area between the conductive parts increased due to the applied mechanical load. Y. Wang et al. [4] developed a dual-frame six-component force sensing system based on a cross-beam elastic element with L-shape supporting beams. Strain analysis and parameter optimization design were performed according to the structure theory and beam theory. A novel calibration machine with two voice coil motors was proposed and decoupling experiments based on the Neural network and the Least Square Method was conducted. Compared with other state-of-the-art force sensors, the measurements up to ± 200 N and ± 15 Nm for force and torque respectively show higher precision (the maximum and mean combination errors are 0.36% and 0.25, respectively). A tri-component force sensing system based on force sensitive resistor array with a mechanically pre-loaded structure was proposed by L. Li et al. [6] for minimally invasive surgery. The proposed system was composed of a stacked structure with four board layers, i.e., force sensitive resistor, reference electrode board, electrode array board, and signal conditioning board. Experimental results indicated that the proposed force sensing system can measure the stiff inclusions and help the articulated robotic probe safely navigate with minimal force resolution of 0.1 N in a measurement range of 0 – 8 N. Disadvantages of this force sensing system always suffers from low accuracy due to its sensing element of force sensitive resistors.

2.2 CAPACITIVE FORCE SENSING SYSTEMS

Another widely used measurement method for multi-component force sensing systems is concerned with the capacitive technique. Multi-component capacitive force sensing systems can detect force and torque components with extremely high resolutions such as mN even pN. Multi-component capacitive force sensing system output can be acquired with switched-capacitor circuits, capacitance-to-frequency converters (oscillators), and capacitive ac-bridges, etc. [5]. As shown in Fig 2.2, parallel-flat configuration is always used in multi-component capacitive force sensing systems.

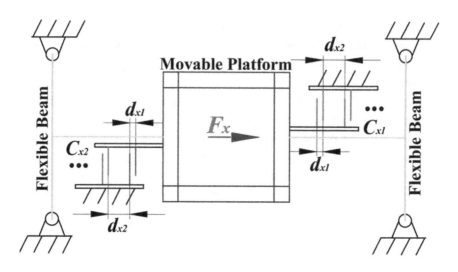

FIGURE 2.2   Force Sensing Systems Based on Capacitive Technique.

The initial capacitances of the parallel plate capacitive sensing system can be expressed as follows:

$$C_{x1} = n\frac{\varepsilon_r\varepsilon_0 ab}{d_{x1}} + n\frac{\varepsilon_r\varepsilon_0 ab}{d_{x2}} + \left(1+\frac{1}{m}\right)\frac{n\varepsilon_r\varepsilon_0 ab}{d_{x1}} \tag{7}$$

$$C_{x2} = C_{x1} \tag{8}$$

where n denotes the number of conducting plate pairs of the capacitive sensor, $a \times b$ is the overlapping area of the conducting plates, d_{x1}, d_{x2} are the initial gap width between electrodes (initial value $md_{x1} = d_{x2}$), ε_r and ε_0 denote the relative and vacuum permittivity, respectively.

Specifically, when applied with force along with the x-axis, the movable platform of the sensing system will deflect with deflection of dx. Assuming that $m \gg 1$, and $d_{x1} \gg x$, the capacitance variation of C_{x1} and C_{x2} can be derived as follows:

$$C'_{x1} = k\frac{\varepsilon_r\varepsilon_0 ab}{d_{x1}-x} = C_{x1}\frac{1}{1-x/d_{x1}} = C_{x1}\left[1+\frac{x}{d_{x1}}+\left(\frac{x}{d_{x1}}\right)^2+\left(\frac{x}{d_{x1}}\right)^3+\cdots\right] \tag{9}$$

$$C_{x2} = k\frac{\varepsilon_r\varepsilon_0 ab}{d_{x1}+x} = C_{x2}\frac{1}{1+x/d_{x1}} = C_{x2}\left[1-\frac{x}{d_{x1}}+\left(\frac{x}{d_{x1}}\right)^2-\left(\frac{x}{d_{x1}}\right)^3+\cdots\right] \quad (10)$$

In order to obtain a more stable and linear reading from the sensing system, a differential measurement configuration have been generally adopted. The relative change of the differential measurement circuit can be described as

$$\Delta C_x = C'_{x1}-C'_{x2} = 2C_{x1}\left[\frac{x}{d_{x1}}+\left(\frac{x}{d_{x1}}\right)^3+\left(\frac{x}{d_{x1}}\right)^5+\ldots\right] \quad (11)$$

The sensitivity of the force sensing systems based on capacitive technique can be defined as follows:

$$\lambda_x = \frac{\Delta C_x}{x} = \frac{C'_{x1}-C'_{x2}}{x} \doteq 2\frac{C_{x1}}{d_{x1}} \quad (12)$$

Additionally, the relative nonlinear error of the force sensing systems based on capacitive technique is given by the following:

$$\delta_x = \left|\frac{2(x/d_{x1})^3}{2(x/d_{x1})}\right| = (x/d_{x1})^2 \times 100\% \quad (13)$$

Finally, the output of the force sensing systems can be derived by a differential capacitive voltage divider circuit with a excitation voltage of V_s.

$$V_{ox} = V_s\left(\frac{C_{x1}-C_{x2}}{C_{x1}+C_{x2}}\right) \quad (14)$$

In summary, multi-component force sensing systems based on the capacitive technique have many advantages, such as high-resolution, simple structure, wide frequency bandwidth, excellent stability, and durability. However, capacitive type sensing systems generally suffer from disadvantages such as stray capacitance, edge effect, parasitic capacitance, and complex with the compulsory electronic circuits.

A simple capacitive force sensing system consisting of a bottom electrode, a flexible central part, and an upper plate with the electrode was proposed by M. Kisić [7]. The sensor was fabricated with 3D printing technology with PLA and silver conductive ink. The measurement sensitivity of the proposed force sensing system is 0.3 pF/N with a maximum detection range of 0.5 N to 3.5 N.

C. H. Na et al. [8] proposed a novel capacitive force sensing system with a dynamic range. The proposed force sensing system mainly consisted of two dielectric layers and a substrate with electrodes. The dielectric layer was composed of two wrinkle-structured elastomer layers, which were arranged in perpendicular direction. The maximum sensitivity of 0.06%/kPa was obtained in the maximum measurement range of 1 MPa.

At the Department of Mechanical Engineering, Southern Methodist University, USA, a composite 3D printed capacitive force sensing system was fabricated via fiber encapsulation and thermoplastic elastomer additive manufacturing technologies [9]. The proposed capacitive force sensing system was composed of a thermoplastic elastomer dielectric spacer and a rigid frame with embedded wires that was fabricated by 3D printing technology. The linear relationship between the applied force and corresponding capacitance variation of the proposed system was obtained with a capacitive sensitivity of about 0.231 pF/kN.

2.3 INDUCTIVE FORCE SENSING SYSTEMS

Inductive sensors also refer to magnetic loop sensors, pickup coil sensors, and search coil sensors. inductive force sensing systems adopt Faraday's law of induction, which measures the changes of self-inductance or mutual inductance of the inductive loop, to detect the applied load.

The generated or induced voltage according to fundamental Faraday's law of electromagnetic induction can be computed as follows:

$$E_m = -n\frac{d\Phi}{dt} = -nA\frac{dB}{dt} = -\lambda_0 nA\frac{dH}{dt} \tag{15}$$

where Φ is the magnetic flux passing through a coil, n in the number of turns with an area A, and B is the external magnetic flux density.

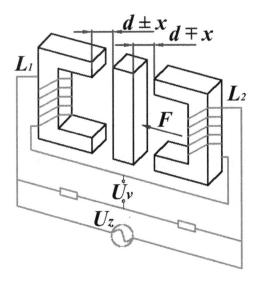

FIGURE 2.3   The Principle Representation of an Inductive Force Sensing System.

The detection principle of an inductive force sensing system with variable inductance is illustrated in Fig 2.3 The system produces a voltage U_v that is proportional to the applied load F.

Specifically, inductive force sensing systems consist of an induction loop with variable inductance, which can be categorized into the differential transformer, reluctance, impedance, and mutual inductance. The transfer function results from the fundamental Faraday's law of induction can be obtained as:

$$\Delta L = L_1 - L_2 \approx 2L\frac{x}{d} \tag{16}$$

$$\dot{U}_z = \frac{\dot{U}_v}{2}\frac{\Delta L}{L} \tag{17}$$

where U_z is the term for generated or induced voltage, U_v is the excitation source, ΔL is the inductance variation due to the gap variation x which produced by the applied force.

Force sensing systems based on the inductive approach generally reveal advantages such as wide dynamic range, the capability of working in harsh environmental conditions, high linear output. However, inductive force sensing systems usually exhibit poor repeatability, poor spatial resolution, rugged mechanical construction, and complicated electronic circuit. Without special design and treatment, inductive force sensing systems are difficult to achieve multi-component detection simultaneously.

2.4 PIEZOELECTRIC FORCE SENSING SYSTEMS

As illustrated in Fig 2.4, force sensing systems based on the piezoelectric approach adopt the piezoelectric effect of quartz crystal or ceramics elements to measure the applied load by converting them to an electrical charge. At present, piezoelectric materials can be divided into three categories: piezoelectric crystals, piezoelectric ceramics, and new types of piezoelectric materials. Among them, piezoelectric crystals are single crystals, such as GaAs; piezoelectric ceramics are polycrystals made of $BaTiO_3$ and other materials. New piezoelectric materials include high-molecular organic piezoelectric materials such as piezoelectric rubber.

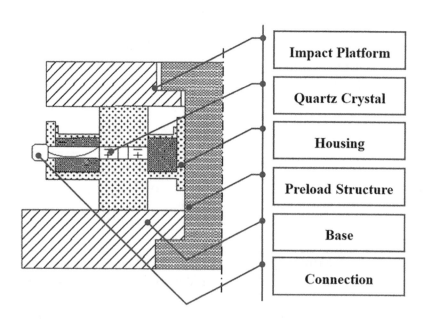

| Impact Platform |
| Quartz Crystal |
| Housing |
| Preload Structure |
| Base |
| Connection |

FIGURE 2.4   Schematic Representation of a Piezoelectric Force Sensing System.

Specifically, the amount of the produced charge can be calculated as follows, which is proportional to the applied load according to the longitudinal piezoelectric phenomenon of piezoelectric sensing material,

$$C_L = \kappa \varsigma F \qquad (18)$$

where κ is the number of mechanically stacked elements, ς is the piezoelectric coefficient, and F is the applied force.

Piezoelectric force sensing systems always show a wider dynamic range due to the high elasticity modulus of piezoelectric materials. Moreover, they also exhibit high frequency response, high accuracy, finer stability, superior reproducibility, and excellent linearity within a wide detection range. However, piezoelectric force sensing systems generally suffer from its inflexibility, requirement of timely recalibration, and incapability of static measurement.

B. C. Rao et al. designed and developed an integrated force sensing system for online cutting geometry inspection [10] based on a piezoelectric force sensor with a high resolution of 0.44 mN and a sensitivity of 7mV/gm. The system can be adopted to detect the feed force combined with thrusts during the diamond-cutting process.

M. Li et al. [11] proposed a piezoelectric force sensing system with high performance for dynamic force detection of landslide. With the capacitive circuit voltage distribution method (CCVDM) and the self-structure pressure distribution method (SSPDM), the compressive capacity is improved and the high direct response voltage is decreased. The sensitivity coefficient of the sensing system was 0.0081 V/kN under different static preload levels from 0 to 500 kN.

A piezoelectric force sensing system was presented in Ref. [12] for atomic force microscopy (AFM). In order to overcome the complexity of the conventional AFM, the elastic element of the sensing system was comprised of a cantilever with a piezoelectric thin film, which was formed on a thermal SiO_2 microbeam. The first resonance frequency and spring constant were 72.5 kHz and 8.7 Nm^{-1}, respectively, and the feasibility of the sensing system for non-contact AFM was verified via experiments.

Y. J. Li et al. [13, 14] proposed a novel six-component piezoelectric force sensing system based on a parallel frame with four point supporting structure for heavy-load manipulators. Piezoelectric quartz was adopted to design the elastic element. The behavior of the elastic element was analyzed by ANSYS, and the calibration experiment was carried out. Experimental results indicated that the proposed sensing system presented high natural frequency, excellent linearity and rigidity, and the measurement showed an interference error of 5%.

An active design approach for no-elastic piezoelectric six-component force sensing system was proposed by J. Liu et al. [15] to address the bottleneck of multi-component force sensing system with an elastic element. The active design approach consists of a numerical simulation model and a static analytical model. Experimental results showed that the proposed design approach was effective for piezoelectric force sensing systems.

A two-component piezoelectric force sensing system was developed in Ref. [16] for a chip refiner. The force sensing system was composed of a probe tip and four piezoelectric elements supported the probe tip. Two-component force sensing system along normal and tangential directions can be detected by any two of the four piezoelectric elements during refining. The proposed force sensing system was revealed the effectivity at low refiner speed. However, it difficult to detect detailed monitoring of the magnitude of the two components of force due to the resonant vibrations.

2.5 FBG-BASED MULTI-COMPONENT FORCE SENSING SYSTEMS

Among the different force detection approaches, force sensing systems based on opto-electric effect, especially the FBG-based force sensing systems, have attracted widespread attention for automatic applications. The principle of FBG-based force sensing systems is to modulate external stress and strain parameters into light wave parameters through the sensing function, and the receiver will demodulate the light wave to obtain the applied load on the sensing system. The FBG-based detection method exhibits excellent characteristics such as high sensitivity, inherent explosion-proof, anti-electromagnetic interference, corrosion resistance, high measurement accuracy (strain measurement accuracy can reach $0.5\text{-}1\mu\varepsilon$), low cost, and electrical insulation. FBG-based force sensing systems have been widely applied in the robotic perception system. For example, fiber Bragg gratings are integrated into the surgical forceps and probes of the robot-assisted surgery system to realize the detection of the operating force in milli-newton level.

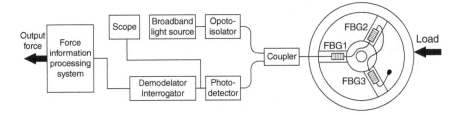

FIGURE 2.5   Schematic Representation of a FBG-Based Force Sensing System.

In addition, fiber Bragg gratings have unidirectional sensitivity (only sensitive to axial strain), which can prevent coupling among components due to the lateral strain during multi-component measurement. Therefore, FBG-based multi-component force sensing system exhibits high sensitivity, good reliability, and strong interference ability, small coupling error, which enables it is suitable for precise measurement in an unknown environment.

The detection principle of the FBG-based force sensing system is shown in Fig 2.5. The fiber grating is made using the photosensitive characteristics of the optical fiber. The optical fiber is used not only as a transmission medium but also as a force-sensitive element for detecting force/torque information.

The wavelength of the reflected signal (λ_b) can be obtain as follows:

$$\lambda_b = 2n_e \Lambda_f \tag{19}$$

where n_e represents the effective refractive index of the grating in the fiber core, and Λ_f is the periodicity of the grating pitch.

the central Bragg wavelength shift of the reflected signal ($\Delta\lambda_b$) due to the applied force or torque can be expressed as

$$\Delta\lambda_b / \lambda_b = (1-p_c)\varepsilon_a + (\zeta_\Lambda + \zeta_n)\Delta T \tag{20}$$

$$p_c = \left(\frac{n_e^2}{2}\right)[\tau_2 - v(\tau_1 + \tau_2)] \tag{21}$$

where p_c is the photo-elastic constant of the Bragg grating, ζ_Λ and ζ_n are the thermal expansion coefficient and the thermo-optic coefficient of the fiber, respectively. ΔT is the temperature change in the optical fiber, v is the Poisson's ratio (0.17), τ_1 and τ_2 are Pockels coefficients of silica (respectively 0.11 and 0.25), respectively.

Fiber–optic sensing technology has appeared as an excellent detection approach from the early 1980s [18]. Optical force sensing systems have shown a great deal of promise, and they have been very successful in physical sensing systems due to their advantages, such as electromagnetic immunity and high resolution [19, 20]. Commonly investigated optical force sensing approaches refer to intensity modulation, phase, polarization, wavelength or transit time of light in the fiber. Generally, fiber optic based force sensing systems can be categorized into intrinsic, extrinsic, or hybrid detection approaches. Specifically, the intrinsic approach uses the optical fibers as the sensing element and is responsible for delivering the light. While the intrinsic method adopts external transducers to measure the applied load. And it applied multi-mode optical fiber to transmit modulated light. The hybrid approaches adopt the optical fiber to carry light in and out of the force sensing systems.

Many multi-component force sensing systems based on opto-electric technology has been implemented in a number of systems, which are typically developed to reinforce the exploitation of more intelligent and automatic applications. Literature such as Ref. [21] – [25] presented comprehensive reviews of sensing systems based on opto-electric approaches. Table 2.5 summarize some typical multi-component force sensing systems based on opto-electric method with comparisons.

Numerical researches related to multi-component force sensing systems based on FBG have been proposed in the past decades, and massive experimental results have been collected. A steadily increasing growth in the design and development of the FBG force sensing system can be noticed in academic research during the past decades. The magnitude growth order about the FBG-based force sensing systems reflects the maturity of FBG-based force sensing systems.

Moreover, other detection principles such as magnetic method, which are capable of mechano-electric, mechano-optic, and mechano-magnetic conversions, can be adopted in multi-component force sensing systems due to their special characters. S. Takenawa [31] proposed a novel three-component force sensing system based on electromagnetic induction with an array of inductors and a permanent magnet. The proposed system can

TABLE 2.5 Typical Multi-Component FBG Force Sensing Systems

Developer	Modulation Type	Characterization Method	Size (mm)	No. of Component	Performance Index	Measurement Range	Application Area
Puangmali [26]	Light intensity	Standard F/T (Force/Torque) sensor (ATI Mini 40)	<20 (in diameter)	3	0.02 N (resolution)	±1.5 N in z-axis, ±3 N in x- and y-axis	Minimally invasive surgical palpation
Guo [27]	Wavelength shift of FBG	Standard weights	Φ 20 × 35.5	3	44.116 pm/N (maximum sensitivity)	10 N in z-axis, ±25 N in x- and y-axis	Robotic manipulation
Liu [28]	Wavelength shift of FBG	Analog Force Gauge, NK-500	Φ 56 × 7	3	0.5136 με/N (maximum sensitivity)	5000 N	Milling force measurement
Roesthuis [29]	Wavelength shift of FBG	CCD cameras	Φ 1 × 172	2	0.74 mm (maximum reconstruction error)	n. a.	Shape reconstruction
Li [30]	Wavelength shift of FBG	Commercial tacho-torquemeter (JN338-500A, Range: -500~500 Nm)	n. a.	2	7.02 pm/N·m (sensitivity)	±20 N·m	Online torque detection
Zhang [31]	Wavelength shift of FBG	A jack with a stress gauge	Φ 50 × 35	1	210 N (resolution)	0 to 3000 N	Heavy load measurement
Lai [32]	Wavelength shift of FBG	Load cell	Φ 4.4 × 6	1	24.28 pm/N (sensitivity)	0 to 40 N	Tendon-sheath mechanisms

NOTE: Φ = diameter; F.S. = Full Scale; n. a. = not applicable; p=10^{-12}, μ=10^{-6}.

measure the change in magnetic flux to determine the resultant force, which can not only detect the three-component force but also distinguish slippage and vibration on contact using the output voltage of the inductors directly.

REFERENCES

1. Liang Q, Zou K, Long J, Jin J, et al. "Multi- Component FBG-Based Force Sensing Systems by Comparison With Other Sensing Technologies: a Review." IEEE Sensors Journal 18.18 (2018): 7345–7357.
2. Liang Q, Zhang D, Wu W, and Zou K. "Methods and Research for Multi-Component Cutting Force Sensing Devices and Approaches in Machining." Sensors 16.11 (2016): 1926.
3. Valdastri P, Roccella S, Beccai L, Cattin E, et al. (2005). Characterization of a Novel Hybrid Silicon Three-Axial Force Sensor. Sensors and Actuators A: Physical, 123, 249–257.
4. Singh, D, Tawk C, Mutlu R, Sariyildiz E, et al. "A 3D Printed Soft Force Sensor for Soft Haptics," 2020 3rd IEEE International Conference on Soft Robotics (RoboSoft), New Haven, CT, USA, 2020, pp. 458–463.
5. Wang Y., Hsu C., Sue C. "Design and Calibration of a Dual-Frame Force and Torque Senso," in IEEE Sensors Journal, in press.
6. Liang, Q., Zhang, D., Coppola, G., Wang, Y., et al. "Multi-Dimensional MEMS/micro Sensor for Force and Moment Sensing: A Review." IEEE Sensors Journal 14.8 (2014): 2643–2657.
7. Li L, Yu B, Yang C, Vagdargi P, et al. Development of an inexpensive tri-axial force sensor for minimally invasive surgery 2017 IEEE/RSJ International Conference on Intelligent Robots and Systems (IROS). IEEE, 2017: 906–913.
8. Kisić, Blaž N, Živanov L and Damnjanović M. "Capacitive force sensor fabricated in additive technology." 2019 42nd International Spring Seminar on Electronics Technology (ISSE). IEEE, 2019.
9. Na, Chan-Hoon, and Kwang-Seok Yun, "Capacitive Force Sensor With Wide Dynamic Range Using Wrinkled Micro Structures as Dielectric Layer." Journal of nanoscience and nanotechnology 19.10 (2019): 6663–6667.
10. Saari Matt, Bin Xia, Bryan Cox, Paul S. Krueger, et al. "Fabrication and Analysis of a Composite 3D Printed Capacitive Force Sensor." 3D Printing and Additive Manufacturing 3.3 (2016): 136–141.
11. Rao B.C., Gao R.X., Friedrich C.R. Integrated Force Measurement for on-Line Cutting Geometry Inspection. IEEE Trans. Instrum. Meas. 1995, 44, 977–980.
12. Li Ming, Wei Cheng, Jianpan Chen, Ruili Xie, et al. "A High Performance Piezoelectric Sensor for Dynamic Force Monitoring of Landslide." Sensors 17.2 (2017): 394.
13. Itoh T, and Suga T. "Development of a Force Sensor for Atomic Force Microscopy Using Piezoelectric Thin Films." Nanotechnology 4.4 (1993): 218.

14. Li Ying-Jun, Sun Bao-Yuan, Jun Zhang, Min Qian, et al. "A Novel Parallel Piezoelectric Six-Axis Heavy force/torque Sensor." Measurement 42.5 (2009): 730–736.
15. Li Ying-Jun, Cong Yang, Wang Gui-Cong, Hui Zhang, et al. "Research on the Parallel Load Sharing Principle of a Novel Self-Decoupled Piezoelectric Six-Dimensional Force Sensor." ISA transactions 70 (2017): 447–457.
16. Liu Jun, Li Min, Qin Lan, and Liu Jingcheng, "Active Design Method for the Static Characteristics of a Piezoelectric Six-Axis force/torque Sensor." Sensors 14.1 (2014): 659–671.
17. Siadat A., Bankes A., Wild P.M., Senger J., et al. "Development of a Piezoelectric Force Sensor for a Chip Refiner." Proceedings of the Institution of Mechanical Engineers, Part E: Journal of Process Mechanical Engineering 217.2 (2003): 133–140.
18. Roriz P., Carvalho L., Frazão O., Santos J. L., Simões J. A., "From Conventional Sensors to Fibre Optic Sensors for Strain and Force Measurements in Biomechanics Applications: a Review," Journal of Biomechanics, 47, pp. 1251–1261, 2014.
19. Zhao, Y., F. Xia, M. Q. Chen, R. J. Tong, and Y. Peng, "Optical Fiber Axial Contact Force Sensor Based on Bubble-Expanded Fabry–Pérot Interferometer," Sensors & Actuators A Physical, 272, 2018.
20. Nagura, C., "Optical force sensor and apparatus using optical force sensor," ed, 2016.
21. Hong C. Y., Zhang Y. F., Zhang M. X, Lai M. G. L., Liu L. Q. "Application of FBG Sensors for Geotechnical Health Monitoring, a Review of Sensor Design, Implementation Methods and Packaging Techniques," Sensors & Actuators A Physical, 244, pp. 184–197, 2016.
22. Barrias A., Casas J. R., Villalba S. "A Review of Distributed Optical Fiber Sensors for Civil Engineering Applications," Sensors, 16, pp. 748, 2016.
23. Caucheteur C, Guo T, Albert J, "Polarization-Assisted Fiber Bragg Grating Sensors: Tutorial and Review," Journal of Lightwave Technology, 35, pp. 3311–3322, 2017.
24. Chen J, Liu B, Zhang H. "Review of Fiber Bragg Grating Sensor Technology," Frontiers of Optoelectronics in China, 4, pp. 204–212, 2011.
25. Dziuda L., "Fiber-Optic Sensors for Monitoring Patient Physiological Parameters: a Review of Applicable Technologies and Relevance to Use During Magnetic Resonance Imaging Procedures," Journal of Biomedical Optics, 20, pp. 010901, 2015.
26. Puangmali P, Liu H, Seneviratne L. D., Dasgupta P, Althoefer K. "Miniature 3-Axis Distal Force Sensor for Minimally Invasive Surgical Palpation," IEEE/ASME Transactions on Mechatronics, 17, pp. 646–656, 2012.
27. Guo Y, Kong J, Liu H, Xiong H, Li G, Qin L. "A Three-Axis Force Fingertip Sensor Based on Fiber Bragg Grating," Sensors & Actuators A Physical, 249, pp. 141–148, 2016.
28. Liu M, Zhang Z, Zhou Z, Peng S, Tan Y. "A New Method Based on Fiber Bragg Grating Sensor for the Milling Force Measurement," Mechatronics, 31, pp. 22–29, 2015.

29. Roesthuis R. J, Kemp M., Dobbelsteen J. J. V. D., Misra S. "Three-Dimensional Needle Shape Reconstruction Using an Array of Fiber Bragg Grating Sensors," IEEE/ASME Transactions on Mechatronics, 19, pp. 1115–1126, 2014.

30. Li, T., C. Shi, Y. Tan, Z. Zhou, "Fiber Bragg Grating Sensing-Based Online Torque Detection on Coupled Bending and Torsional Vibration of Rotating Shaft," IEEE Sensors Journal, 17, pp. 1999–2007, 2017.

31. Zhang W, Li E, Xi J, Chicharo J, and Dong, X. "Novel Temperature-Independent FBG-Type Force Sensor." Measurement Science & Technology 16.5 (2005): 1600.2005.

32. Lai, Wenjie, Lin Cao, Zhilin Xu, Phuoc Thien Phan, et al. "Distal end force sensing with optical fiber bragg gratings for tendon-sheath mechanisms in flexible endoscopic robots." 2018 IEEE International Conference on Robotics and Automation (ICRA). IEEE, 2018.

33. Takenawa, S., "A soft three-axis tactile sensor based on electromagnetic induction," in Proc. IEEE International Conference on Mechatronics, 2009, pp. 1–6.

CHAPTER 3

FBG-Based Optical Force Sensing Systems

3.1 ADVANTAGES OF FBG-BASED OPTICAL FORCE SENSING SYSTEMS

With the gradual improvement of fiber grating technology, a variety of fiber grating-based devices and sensors continue to emerge. As a new type of strain sensitive element, fiber gratings are gradually used to develop new types of force sensors. Generally, Opto-electric force sensing systems can be ranked based on the physical quantity being analyzed, among which the most regularly examined measurands are intensity, wavelength, phase, and polarization. FBGs offer significant potential for force measurement applications. The importance of FBG-based optical force sensing systems stems from the following characters compared with the traditional force-sensing systems:

1. Excellent environmental resistance capability. Because there are no electrical connections or current flows through the fiber optical cable, FBG-based optical force sensing systems are undisturbed by electrical noise, and not affected by lighting, electromagnetic, and electrostatic interference.

2. Especially high sensitivity. Generally, FBG-based optical force-sensing systems are capable of providing extremely slight strain changes, i.e., less than 1 $\mu\varepsilon$ (10^{-6}mm/mm), and then detect force and

moment with more excellent sensitivity by sensitivity enhancement methods.

3. Small size and lightweight. The profile of FBG sensing element ranges from a few tens to hundreds of micro-meters with a typical core diameter value around Φ 60 – 200 μm, which can enable the miniaturization and compactness of the force sensing system.

4. Multiplexing capability and compacter system. More than 100 FBGs can be multiplexed on a single optical fiber with a sophisticated interrogation instrument, and even be arranged into arrays because an individual FBG only reflects a particular wavelength of light while transmits others.

Advantages and disadvantages of the FBG-based force sensing system in comparison to other measurement techniques are listed in Table 3.1.

3.2 THE PRINCIPLE OF FBG-BASED OPTICAL FORCE SENSING SYSTEMS

The reflection spectrum type (central wavelength, bandwidth, etc.) of the fiber grating depends on the fiber grating period Λ and the effective refractive index n_{eff}. Any physical process that changes these two parameters will cause the spectrum type of the fiber grating to change. The external force acts on the sensor, causing the FBG attached to the elastic body to strain, which causes the FBG wavelength to change. The change in wavelength is mapped to a change in force by decoupling. Analyzing the sensing principle of fiber grating is of great significance to the further development of a new type of fiber grating force sensor.

Due to the periodic change of the refractive index of the fiber core region, the waveguide conditions of the fiber are changed, which leads to the corresponding mode coupling of the specific wavelength, and the transmission spectrum, as well as reflection spectrum of the wavelength, turn singular.

The measurement principle of the FBG-based optical force sensing system is illustrated in Fig 3.1. In order to detect multi-component force and moment, several FBGs are multiplexed along the length of a fiber and bonded onto the elastic element of a force sensing system. The specific narrow optical reflection-band range at the Bragg-wavelength λ_b of the broadband light source is reflected by every single FBG. Particularly,

TABLE 3.1   FBG Force Sensing System Characteristics by Comparing to Conventional Systems [1]

Sensing Approach	Transduction Description	Advantage	Disadvantage
FBG-based system	refractive index/light intensity/ spectrum variation due to a mechanical force or moment	• good reliability and high sensitivity • wide measurement range • noncontact and high-temperature performance • nonelectrical and immunity to electromagnetic interference • light weight and small size • intrinsically safe in the explosive environments • distributed measurement	• hard to construct dense arrays • more costly • complex information processing system
Resistive	semiconductors conductivity/resistance variation due to a mechanical force or moment	• simple construction and low cost • high and adjustable resolution • high reliability and maintenance-free • compatibility with VLSI	• higher power consumption • rigid and fragile • scarce reproducibility • electromagnetic compatibility • narrow frequency bandwidth • contradictions between flexibility and sensitivity
Capacitive	capacitance variation due to a mechanical force or moment	• high sensitivity and resolution • large bandwidth • robustness • long-time stability • drift-free • durability • adaptability to environment	• complex with the compulsory electronic circuits • stray capacitance • edge effect • temperature sensitivity
Inductive	magnetic coupling variation due to a mechanical force or moment	• linear output • high power output • wide dynamic range	• lower frequency response • poor reliability • massive size
Piezoelectric	generation of a surface charge due to a mechanical force or moment	• high-frequency response • higher accuracy and finer resolution • high sensitivity and superior stiffness • high dynamic range	• charge leakages • require to be embedded into the machining structure • deteriorations of voltages or drifts in the presence of static forces • poor spatial resolution

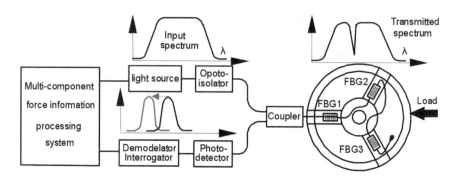

FIGURE 3.1   Functional Diagram of FBG-Based Optical Force Sensing Systems.

the wavelength of this reflection band depends on the strain induced by the applied load. Therefore, the measurand (force and moment) can be detected by monitoring the shift in wavelength of the reflected light.

3.3 DESIGN AND CONSTRUCTION OF A FBG-BASED OPTICAL FORCE SENSING SYSTEM

The whole FBG-based optical force-sensing system consists of a novel elastic element with FBGs, a base, and an upper cover. The base and upper cover are used to equip the sensing system between the robot effector and the end of the robot. The sensing system is arranged along the Z-axis. There are many threaded and through holes on the sensor for the assembly purpose. Specifically, the base is designed with adapted with 8 × M6 through holes, which enable the sensing system to fix to the robot end. In addition, the 6 threaded holes of M3 on the bottom flange can be implemented to mount the elastic element. In the upper cover, there are 4 × M3 through holes for mounting the sensing system onto the effector. The elastic element includes a ring column, three flexible beams, and a cylinder in the middle. For the ring column, its outer diameter, inner diameter, and height are R1 = 74 mm, R2 = 56 mm and, h1 = 5 mm, respectively. For the elastic beam, its length, width, and height are L = 20 mm, d = 5 mm and h5 = 2 mm, respectively. These three elastic beams are evenly distributed within the ring column with an angle =120°. The distance between the upper surface of the elastic beam and the top surface of the ring column is h4 = 1.5 mm. For the inner cylinder, its inner radius, height and the distance from its upper surface to the top surface of the ring column are R3 = 16 mm, h3 = 7.5 mm, and h2 = 3 mm, respectively. It should be noted that there is an overload protection distance P designed in the negative direction of Z-axis, which prevents the robot arm from being damaged by excessive loads when it is engaged in contact tasks. This overload protection distance is set at 0.2 mm.

3.3.1 Structure of the Elastic Element

A novel elastic element with three flexible beams (beam 1, 2, and 3) in the wheelset structure was designed, as shown in Fig 3.1. The flexible beam 1 was arranged along with the X-axis direction, and three FBGs were bonded along the centerlines of the flexible beams. To investigate the stress and strain characteristics of the designed sensing system, a commercial finite element software ANSYS is utilized to conduct the static and dynamic analysis. Since the elastic element is the sensing element of the system, only the elastic element is analyzed, which is made of aluminum alloy with the main material parameters shown in Table 3.2. In these simulations, the lower surface of the ring column is fixed and then the load is applied through the middle cylinder of the elastomer.

3.3.2 Strain Distribution on the Elastic Element

The strain distribution on the elastic element can be used to design the FBG arrangement strategy. To acquire the strain behavior of the elastic element, it is applied with several loads such as $Fx = 100$ N, $Fy = 100$ N, and $Fz = 100$ N, respectively.

The origin of the Abscissa lies in the intersection point between the centerline of the beam and the inner surface of the elastomer ring column, which extends along the beam to the outer surface of the middle cylinder with the length being 20 mm. When $Fx = 100$ N, the centerline strain of Beam 3 is almost zero, but the centerlines of Beam 1 and Beam 2 have been induced large strains, which are symmetrical with respect to each other. In this scenario, the maximum stress of the centerline of the three

TABLE 3.2   Material Parameters of the Elastic Element

Property	Value	Unit
Density	2770	kg/m³
Young's Modulus	7.1 e10	Pa
Poisson's Ratio	0.33	-
Bulk Modulus	6.96 e10	Pa
Shear Modulus	2.67 e10	Pa
Tensile yield strength	250	MPa
Compressive yield strength	250	MPa

beams is 557.64 µε, which is called Maxstrain with its position marked by the green circle. When $Fy = 100$ N, the strain produced in the centerline of Beam 3 shows an increasing trend, while the strains in the centerlines of Beam 1 and Beam 2 decrease with an identical pattern. In this case, Maxstrain = 508.3 µε. When $Fz = 100$ N, the centerline strain of Beam 1, Beam 2, and Beam 3 have shown an overall increasing trend. In this scenario, *Maxstrain* = 1579.80 µε. It can be clearly observed that the strains on the centerlines of the elastic beams are non-uniform. Ideally, ring columns do not generate any strain and deformation.

3.3.3 Decoupling of the Multi-Component Force Sensing System

For the 3-axis force sensor, the calibration data needs to be decoupled for the character analysis in each direction. Consider N calibration data and the goal is to find the mapping between wavelength shift and the applied load. The Extreme Learning Machine (ELM) is implemented as the nonlinear decoupling algorithm to find the relation matrix. It is worth noting that ELM has demonstrated its suitability for multi-component force sensing system sensors and better performance than the commonly used least square method, SVM and MP, in our previous work. The specific decoupling steps of ELM are as follows:

Step 1: calibrate the force sensor and obtain the experimental data $(\Delta\lambda,\tau)\_i$, i=1, ... N. Refer to Section 4.1 for specific calibration steps.

Step 2: adopt the temperature compensation method, and then take the result as the network input of ELM.

Step 3: divide the data set into a training set and a test set.

Step 4: set the value range of hidden layer nodes with $l = 0$, ..., L.

Step 5: $l=l+1$.

Step 6: randomly initialize the weight W and b bias of the hidden layer.

Step 7: calculate the output layer weight β.

Step 8: calculate the mean square error (MSE) of the predicted results.

Step 9: if $l=L$, go to step 10. Otherwise, go back to step 5.

Step 10: The min(MSE) is searched and the corresponding parameters l, W, b and β are saved for predicting the force.

3.3.4 Experimental Results

To obtain the relationship between the FBG strain and the applied load, the designed force sensing system needs to be calibrated by the calibration system. In the calibration process, the elastic element of the sensor is firstly fixed on the base and the loading cap is attached to the sensor. Then, the standard weight of 500 g is tied to the wire rope and connected to the loading cap. The pulley, mounted on a bracket with adjustable height is used to support the wire rope. After that, FBG is connected to the interrogator, which can record the wavelength changes of the four FBGs in real-time. It should be mentioned that the frequency, accuracy, and resolution of the acquisition are 100 Hz, 5 pm and 1 pm, respectively. In the end, the signal is sent back to the computer.

For the calibration of Fx, align the negative X direction of the sensor to the position of the pulley, load a group of loadings from 0 N to 30 N with a step size of 5 N, and then get unloaded with a step size of 5 N. The positive X direction calibration of the sensor is also experiencing the loading and unloading process in the same way. During the calibration, this process has been carried out six times. It should be mentioned that the calibration of Fy is the same as that of Fx. The average characteristic curves of the calibration results for Fx and Fy can be obtained by the above process, respectively.

For the calibration of Fz, a group of loadings from 0 N to 40 N have been loaded with an increase of 5 N in each step, and then unloaded in a similar manner. This process also has been repeated 6 times. The output wavelength changes of the FBG have been recorded and the average characteristic curves of the calibration results can be obtained.

It can be observed that under the loadings of Fx, Fy, and Fz, the trend of the FBG wavelength change is consistent with the simulation results. Specifically, when Fx is loaded, the wavelength change of FBG1 is very small. However, the wavelength change of FBG2 and FBG3, linearly increases and decreases, respectively. When Fy is loaded, the wavelength change of FBG1 increases linearly, while the wavelength change of FBG2 and FBG3 decreases linearly. When Fz is loaded, the wavelength change of FBG1, FBG2, and FBG3 increases linearly. FBG4 is implemented for temperature compensation, which makes the corresponding wavelength change is close to 0 at room temperature.

From the calibration data, the sensitivity and repeatability error of each FBG of the sensing system to Fx, Fy, and Fz can be determined with the corresponding results.

TABLE 3.3 The Sensitivity and Repeatability Errors of FBG1, FBG2, and FBG3 on the Force Sensor

	FBG1		FBG2		FBG3	
	Sensitivity (pm/N)	δ(%)	Sensitivity (pm/N)	δ(%)	Sensitivity (pm/N)	δ(%)
Fx	7.67	1.33	7.81	1.96	0.7386	4.38
Fy	5.45	1.12	3.92	3.04	7	2.2
Fz	3.09	0.79	2.76	0.7	2.52	0.6

As shown in Table 3.3, when Fx is loaded, the maximum sensitivity obtained is 7.81 pm/N in FBG2, and the maximum repeatability error is 4.38% in FBG3; when Fy is loaded, the maximum sensitivity is 7 pm/N in FBG3, and the maximum repeatability error is 3.04% in FBG2; when Fz is loaded, the maximum sensitivity is 3.09 pm/N in FBG1, and the maximum repeatability error is 0.79% in FBG1.

3.4 STATE OF THE ART OF THE FBG-BASED OPTICAL FORCE SENSING SYSTEMS

3.4.1 FBG-Based Force Sensing Systems with Single Component

There is an increasing trend of establishing FBGs in various single-component force sensing applications with longitudinal strain measurement technology. Lai et al. [2] designed a novel FBG-based force sensing system for flexible endoscopic surgical robots. A high sensitivity of 34.14 pm/N and a relatively low measurement error of 0.178 N were achieved by embedding a 1 mm FBG to a nitinol tube (3 mm long). Experimental results showed that the force sensing system was capable of sensing haptic force in TSM-driven surgical robots. It was also demonstrated that the proposed sensing system can be used in both TSM- and tendon- driven robots.

A new fiber optic load sensing system was proposed by Liu et al. [3] with Long-Period Fiber Gratings based on the transverse strain-introduced birefringence measurement technology. Experimental results show that the proposed sensing system achieved high transverse strain sensitivity of 500 nm/kg/mm, and offered possible applications for multiple component force measurement.

Abushagur et al. [4] demonstrated a 1D FBG-based force sensing system for microsurgical instruments to address the cross-talk between the axial and transversal forces. A flexure elastic element was designed and a tapered FBG was bonded onto that. Calibration and temperature compensation experiments were carried out, and their results indicated that the system can detect axial force information with an RMS error of 0.356 N with a full scale of (10 N).

Lim et al. [5] proposed a force sensing system based on FBG technology to measure the grasping force for teleoperated surgical robots. An FBG was bonded onto the grasper, which enables a force sensitive forceps with high sensitivity and resolution of 11 mN.

Shi et al. [6] proposed a high-sensitivity force sensing system based on FBG with a relatively large measurement range for laparoscopic surgery. A novel elastic element with suspended optical fiber based on the compliant Stewart structure and a force-sensitive flexure was proposed. Calibration experiments demonstrated that the sensing system can actualize a high resolution of 21 mN within a wide measurement range.

A miniature FBG-based force sensing system with a miniature elastic element for tissue palpation during minimally invasive surgery by Lv et al. [7]. Structural optimization based on simulation was implemented to increase the linearity and resolution performance, and a small linearity error of 0.97% as well as high resolution of 2.55mN were achieved (0~5N measurement range). In vitro palpation implementation and ex vivo indentation experiments validated the effectiveness of the proposed system.

A novel approach for detecting cutting force based on FBG sensor was proposed by Jin et al. [8]. Competitive advantages such as sensitivity of 2.51 mV/N, linearity error of 1.2% F.S., and natural frequency of 950 Hz were proved by manufacturing test and calibration experiments.

Bartow et al. [9] demonstrated that FBGs can be applied to dynamic force monitoring for optoelectronic micro machine tool. The FBG sensor is mounted on the surface of the tool, and the movement of the tool tip is accurately evaluated by comparison with a piezoelectric accelerometer. The cutting test results proved the feasibility of the system.

A high-sensitivity strain sensor based on FBG and flexible hinge was proposed by Liu et al. [10] for detecting the structure surface strain. A flexible hinge bridge displacement magnification structure was designed, and the strain sensitization amplification factor was obtained through energy theory and flexible matrix approach. A calibration experiment was carried out, and the results indicated that the proposed system was characterized by a relatively high sensitivity of 10.84 pm/μ and can be adopted to monitor dynamic strain and force.

3.4.2 FBG-Based Force Sensing Systems with Multiple Components

Most single-component FBG-based force sensing systems were developed with longitudinal strain measurement method. With the advantage of distributed measurement, a multi-parameter fiber grating system is capable of monitoring forces along multiple axes. Actually, FBG-based force sensing systems can predict forces along 2 directions of the plane with transverse and longitudinal strain measurement capability. Moreover, fiber grating systems are capable of detecting forces and torques along multiple axes with the advantage of distributed measurement.

Li et al. [11] proposed a three-component force sensing system based on the FBG technology for aspiration instruments in neurosurgery. Specifically, the sensing system can detect axial force Fz and torques Mx and My through a novel elastic element with an annular diaphragm. The theoretical model and decoupling principle of the proposed sensing system were obtained, and the calibration and temperature compensation experiments were carried out. Experimental results showed that the proposed sensing system could measure three component force and torques with a high resolution of 8.8 mN and a relative error of 4.4% of the full scale (2 N).

Most force sensors based on one-dimensional FBG are developed using longitudinal strain measurement methods. However, FBG sensors can simultaneously measure lateral and longitudinal strains [12]. The capability of lateral strain measurement enables FBG-based sensing systems to detect forces along two components [13] – [16].

Gonenc et al. [17] designed an approach of using an integrated fiber Bragg grating strain sensor to predict the lateral and axial forces of retinal microsurgery. Specifically, three lateral FBGs were evenly fixed around the catheter to detect the lateral force, while the fourth FBG was located inside the catheter to predict the axial force. Calibration and verification experimental results indicated that the method can predict lateral force and axial force, with root mean square (RMS) errors of 0.15 mN and 1.69 mN, respectively.

A novel 3-component FBG-based force sensing system was developed for a robot foot with a Maltese cross beam by Xiong et al. [18]. The 3-component force information that was ranging in ± 50N, ± 50N, and 0-60N, respectively, was acquired by the system with 5 FBGs of 3-nm wavelength interval. Calibration experimental was performed, and its results indicated that the repeatability and coupling error of the system are less than 2.87 N and 4.045 N, respectively. However, there is no actual force measurement experiment carried on an actual robot foot.

Recently, Choi et al. [19] described a 3D force sensing system based on FBG technology, which can measure three directional forces with flexible titanium structure and three FBGs. The proposed sensor has been verified by experiments, and the measurement range is – 12 N to 12 N. Thanks to the calibration and compensation method based on artificial neural network, the maximum error and sensitivity of the multiple component force estimated by the system are 0.12 N and 0.06 N, respectively.

Müller et al. [20] proposed a method of detecting multiple component force and torque using an optical fiber containing six 3 mm long FBGs. Calibration and text experimental results showed that the proposed method had competitive advantages. The condition numbers of the normalized compliance matrix and the maximum force resolution were 28.9 and 100 mn, respectively.

A cutting force sensing system based on FBG approach with a novel elastic element characterized by four rings in annulus shape was proposed by Liu et al. [21] for milling operations. Experimental results indicated that the system maximum detection error was around 15.6 N within the maximum measurement range of 400 N, which verified that the proposed system can be used for turning and milling processing.

Kim et al. [22] proposed a novel 6 component force sensing system based on FBG for haptic feedback of loadings at end-effectors of a minimally invasive surgical robot. FBG was bonded on a novel platform frame with a pretension of 2000 μm strain. Finite element analysis was carried out to obtain the corresponding strain. The experimental results correlated well with that of FEA with 3% ~ 4% error.

3.5 TARGET APPLICATIONS

The optical force sensing technology has been approaching maturity in the past 30 years [23] – [26]. Early engineers and scientists such as Kane noticed great potential of the sensing method based on fiber optical in military and industrial applications [27]. For example, Hirose et al. [24] proposed a novel multi-component optical force sensing system based on light intensity detection method. Meanwhile, many investigators adopted the optical force sensing technology to practical applications such as medical systems [28] – [29], robot control [30] – [32], special detection applications under special environments (for example, Electro-Magnetic Interference environments [33]). Some investigations have been carried out with FBG-based optical force sensing systems, and many of which have already arrived commercialization phases [34] – [38]. However, continuous performance improvement is still challenging, and massive efforts have been made in order to achieve more cost-effective and high-performance approaches as well as exploring new applications.

3.5.1 Robot Applications

Precise prediction of multi-component force has been becoming more and more important and received massive attention in a number of robot applications [39]. However, force sensing systems are still hard to be equipped in real automation systems.

Numerous developers revealed their important contribution to force sensing technologies for robot applications. Recently, Guo et al. [40] proposed a three-component force sensing system based on FBG for a robot fingertip. The sensing system can detect the three-component force with FBGs for robot fingers. Specifically, six FBGs were bonded onto an elastic force-sensing element. The wavelength shift of the FBGs due to the applied load was recorded by an interrogator. Calibration experiment results indicated that the sensing system can detect the multi-component force with a minimum precision of 0.061 N, 0.059 N, and 0.045 N for Fx, Fy and Fz, respectively.

Melchiorri et al. [41] developed a novel three-component force sensing system based on optoelectronic approach for robot fingers. The applied load was predicted by measuring the scattered energy density variation with photodetectors. Static and dynamic calibration experimental results showed that the proposed force sensing system was effective and can locate the contact point position. Recently, the author and his team extended the proposed optical technology to a six-component force sensing system for robot fingertips and grippers [42]. A calibration experiment was carried out with an ATI sensor (SI-130-10). Experimental results showed that the system can detect six-component force with a maximum relative error of 10% of the maximum measurement range ([-50, 50] N and [-1, 1] Nm for force and torque components, respectively).

A Magnetic Resonance Imaging-guided needle placement robot was developed by Su et al. [43] from Worcester Polytechnic Institute for real-time in situ needle steering applications. Based on fiber optic measurement technology, the sensing system was integrated with a Fabry-Perot interferometer and a 16-bit data acquisition system. It was confirmed that the force sensing system was effective with a gauge factor of 47.48 mv/με in MRI compatibility test and experiments.

3.5.2 Force Sensing in Special Environments

Special environments such as magnetic resonance imaging application scenarios also need force sensing systems. Recently, Tan et al. [44] proposed a novel MRI-compatible force sensing system based on optical technology for continuous MR imaging. A novel elastic element with a shape of elastic frame structures fabricated with polymer materials was proposed. The topology optimization method was adopted to design the structure design with the goal of maximum resolution and bandwidth performance. In order to account for polymer material hysteretic behavior, the Prandtl-Ishlinskii (PI) play operator was applied. The experimental results indicated the force sensing system based on optical approach was feasible with a resolution of 3 mV and a maximum RMS error of 0.525 N.

Similarly, Zhang et al. [45] designed a force sensing system based on the optical approach with an excellent temperature-independent characteristic. With a sophisticated elastic element with a spoke-type structure and two FBGs with different wavelengths, the sensing system can measure the applied load with a high resolution of 0.21 kN in a maximum measurement range of 0 to 3 kN. The temperature-independent characteristic enabled the sensing system can automatically compensate the temperature cross-interference effect between −20 °C and 75 °C with differential detection technologies.

3.5.3 Biomedical and Medical Applications

Minimally invasive surgery (MIS) has been innovated due to the innovative development and upgrading of instruments and specific technologies because it has notable advantages over traditional open surgery, and its wide application is limited by lack of force and haptic feedback. Many studies have been proposed to support force and tactile feedback for biomedical and medical applications [46], [47]. However, due to the regulation of clinical implementation, research and development results are usually not used in actual devices or systems.

Puangmali et al. [48] proposed a precision intensity modulated fiber optic force sensing system, which has three force components and was used for minimally invasive surgical palpation. The three-component force information provided by the sensor is used to identify changes in tissue stiffness, thereby locating tissue lesions, such as tumors. The hysteresis error of the sensor was 3.5%, and the root mean square error was 0.68%. The proposed force sensing system together with an ATI Mini 40 force sensor were installed in the robot platform for verification. Experimental results on pig kidneys show that the proposed sensing system can detect and locate invisible tumors.

In order to achieve multi-component force prediction in minimally invasive robotic surgery, Haslinger et al. [49], German Aerospace Center (DLR) developed a new type of force sensing system based on the optical fiber method. Taking the Stewart platform as the elastic element, the smallest 6-component FBG force sensing system (Φ6.4mm×6.5mm) was proposed. A calibration experiment was carried out, and the results showed that the maximum error of crosstalk between components was 6.5657 Nmm, and the maximum hysteresis error was 0.5320 N.

In order to be applied in cardiac catheterization guided by MRI, polygerinos et al. [50], [51] studied the sensing method based on FBG, and specially designed the catheter tip based on light intensity modulation technology. In order to measure the applied load, three plastic optical fibers were integrated into the tip of the catheter in a circular pattern, and the reflector was bonded to the tip of the catheter with a special elastic material. The optical intensity of the optical fiber was modulated by the distance between the reflecting surface of the reflector and the aligned optical fiber. Calibration was conducted using a commercial force sensor (ATI nano 17) and a linear micrometer calibration tool. In addition, experiments were carried out under laboratory conditions and in vivo real-time MRI. Experimental results show that the proposed system can predict the force with a resolution of 0.0478 N and small hysteresis in the maximum measurement range of 0 – 0.85 N.

A multi-component force sensing system was designed and developed by Gonenc et al. [52] based on photoelectric methods for robot-assisted vitreoretinal surgery. The system uses four embedded fiber grating sensors, linear regression, and nonlinear fitting method based on the second-order Bernstein polynomial, which can accurately predict the tensile force component along the tool axis in a measurement range of 0-25 mN. Calibration and verification experiments were conducted, and experimental results indicated that the system can predict the three-component force, and the root mean square error is below 0.15mn and 2mn, respectively.

3.5.4 Dynamic Measurements

In terms of dynamic force and torque measurement [53], the sensing technology based on FBG has shown significant advantages.

Li et al. [54] recently studied a novel dynamic torque detection system based on FBG sensing approach. In order to acquire bending and torsion information, two FBG were bonded onto the shaft with the opposite position. Real-time detection experiments were conducted, and results indicated that the proposed system was effective for dynamic vibration prediction with a sensitivity of 7.02 PM/nm for rotating machinery at 400 Revolutions per minute (RPM). In addition, the experimental results at different speeds show that the proposed system can measure the applied torque at a speed of 400 RPM (revolutions per minute), and its accuracy is higher than that of the commercial tachometer torque meter. The advantages of this system make it possible to measure the vibration of rotating machinery.

A novel FBG-based force sensing system was proposed by Roesthuis et al. [55] for manipulating multi segment continuum manipulators. Through a nickel titanium alloy wire, 12 FBGs were fabricated into 3 optical fibers and introduced into the hollow framework of the continuous manipulator. An interrogator (Deminsys Python) is used to connect optical fiber and detect the variation of reflection wavelength caused by strain. The system expressed the possibility of simultaneous measuring interaction force and bending.

Park et al. [56] have described a tactile interaction system based on force measurement technology of micro fiber. Three optical fibers with FBGs were bonded into the groove of the probe to realize the real-time estimation of the needle tip position and deflection. The strain and temperature occurred on the needle can be measured by grating in different directions and positions. Two high-resolution digital cameras are used for calibration experiments. The calibration and experimental results show that for 6.8 mm two-point bending deflection, the system can detect the deflection and bending shape with an error of 0.6 mm.

3.5.5 Other Measurement Applications

a new optical fiber force sensing system was proposed by Malla et al. [57] for detecting the wheel load of vehicles on the highway. The system can be used to estimate the magnitude and position n (850 pounds) of force in the measurement range of 3781 by using a fiber with two concentric photoconductive regions as sensing elements and a forward time division multiplexing (FTDM) method. Experiments on Commercial General Test loaders (SATECTM Series) as well as wheel load tests were carried out.

The results show that the system can predict the wheel load in the moving load (WIM) system with a sensitivity of 2.1×10^{-3} mV/n. The proposed system has advantages such as weighing the vehicle while driving rather than at rest, so it is economically feasible to carry out measurement tasks on busy roads.

An exoskeletal force sensing system with 5 FBGs approach was designed by Park et al. [58] for composite end-effectors. A miniaturized polyurethane finger with a hollow shell exoskeletal structure (15 mm) was fabricated via shape deposition manufacturing technology. Experimental results verified the capability of detecting the contact location as well as multi-component force with a 0.15 N resolution in a measurement range of 5 N.

3.6 CHALLENGES AND FUTURE DIRECTIONS

Notable successes in the fiber optic sensing technology have been achieved. However, force sensing system based on FBG still confronted difficulties in real industrial applications. For example, sterilizability for medical applications, necessary to enhance its resistance to enlarge the measurement range of the sensing system, sensitization treatment for high-precision monitoring applications, and expensive optical spectrum monitoring systems, etc.

REFERENCES

1. Liang Q, Zou K, Long J, Jin J, et al. "Multi-Component FBG-Based Force Sensing Systems by Comparison With Other Sensing Technologies: a Review." IEEE Sensors Journal 18.18 (2018): 7345–7357.
2. Lai W, Cao L, Tan RX, Phan PT, et al. "Force Sensing With 1 Mm Fiber Bragg Gratings for Flexible Endoscopic Surgical Robots." IEEE/ASME Transactions on Mechatronics 25.1 (2019): 371–382.
3. Liu Y., Zhang L, Bennion I, "Fibre Optic Load Sensors With High Transverse Strain Sensitivity Based on Long-Period Gratings in B/Ge Co-Doped Fibre," *Electronics Letters*, vol. 35, pp. 661–663, 2002.
4. Abushagur A. A. G., Arsad N, Elgaud M. M., Bakar A. A. A. "Development of a 1-DOF Force Sensor Prototype Incorporating Tapered Fiber Bragg Grating for Microsurgical Instruments," IEEE Access, 7, pp. 168520–168526, 2019.
5. Lim SC, Lee HK, Park J. "Grip Force Measurement of Forceps With Fibre Bragg Grating Sensors," Electronics Letters, 50, pp. 733–735, 2014.
6. Shi C, Li M, C. Lv, J. Li and S. Wang. "A High-Sensitivity Fiber Bragg Grating-Based Distal Force Sensor for Laparoscopic Surgery." IEEE Sensors Journal 20.5 (2020): 2467, 1 March1, 2020.

7. Lv C., Wang S., and Shi C. "A High-Precision and Miniature Fiber Bragg Grating-Based Force Sensor for Tissue Palpation During Minimally Invasive Surgery." Annals of Biomedical Engineering 48.2 (2020): 669–681.

8. Jin W. L., Venuvinod P. K., Wang X, Jin W. L., Venuvinod P. K., Wang X. "An Optical Fibre Sensor Based Cutting Force Measuring Device," Transactions of Nanjing University of Aeronautics and Astronautics, 35, pp. 114–117, 1994.

9. Bartow M. J., Calvert S. G., Bayly P. V. "Fiber Bragg grating sensors for dynamic machining applications," *in Pacific Northwest Fiber Optic Sensor Workshop*, 2003.

10. Liu M, Wang W, Song H, Zhou S, et al. "A High Sensitivity FBG Strain Sensor Based on Flexible Hinge." Sensors 19.8 (2019): 1931.

11. Li T., King Nicolas Kon Kam, and Ren Hongliang. "Disposable FBG-Based Tridirectional Force/Torque Sensor for Aspiration Instruments in Neurosurgery." IEEE Transactions on Industrial Electronics 67.4 (2019): 3236–3247.

12. Liu Y, Chiang KS, Chu PL "Fiber-Bragg-Grating Force Sensor Based on a Wavelength-Switched Self-Seeded Fabry-Pe/spl acute/rot Laser Diode," *IEEE Photonics Technology Letters*, vol. 17, pp. 450–452, 2005.

13. Udd, E, Schulz WL, Seim JM, Trego A, et al. "Transversely Loaded Fiber Optic Grating Strain Sensors for Aerospace Applications," *in Spie's International Symposium on Nondestructive Evaluation and Health Monitoring of Aging Infrastructure*, 2000.

14. Haran FM, Rew JK, Foote PD., "Fiber Bragg Grating Strain Gauge Rosette With Temperature Compensation," Proceedings of SPIE - The International Society for Optical Engineering, 3330, pp. 220–230, 1998.

15. Magne S, Rougeault S, Vilela M, Ferdinand P "State-of-Strain Evaluation With Fiber Bragg Grating Rosettes: Application to Discrimination between Strain and Temperature Effects in Fiber Sensors," *Applied Optics*, 36, pp. 9437–47, 1997.

16. Silva-López M, MacPherson WN, Li C, Moore AJ, et al., "Transverse Load and Orientation Measurement With multicore Fiber Bragg Gratings," *Applied Optics*, 44, pp. 6890, 2005.

17. Gonenc B, Iordachita I, "FBG-Based Transverse and Axial Force-Sensing Micro-Forceps for Retinal Microsurgery," *Sensors*, 2017, pp. 1–3.

18. Xiong L., Jiang G., Guo Y., Liu H., "A Three-Dimensional Fiber Bragg Grating Force Sensor for Robot," *IEEE Sensors Journal*, 18, pp. 3632–3639, 2018.

19. Choi H., Lim Y., Kim J. "Three-axis force sensor with fiber Bragg grating," *39th Annual International Conference of the IEEE Engineering in Medicine and Biology Society, Seogwipo, South Korea*, 2017, pp. 3940–3943.

20. Müller MS, Hoffmann L, Buck TC, Koch AW, "Fiber Bragg Grating-Based Force-Torque Sensor with Six Degrees of Freedom," *International Journal of Optomechatronics*, 3, pp. 201–214, 2009.

21. Liu M., Zhou Z, Tao X, Tan Y. "A Dynamometer Design and Analysis for Measurement the Cutting Forces on Turning Based on Optical Fiber Bragg Grating Sensor," *Intelligent Control and Automation*, 2012, pp. 4287–4290.

22. Kim Cheol, and Lee Chan-Hee. "Development of a 6-DoF FBG Force–Moment Sensor for a Haptic Interface with Minimally Invasive Robotic Surgery." Journal of Mechanical Science and Technology 30.8 (2016): 3705–3712.

23. Kersey A. D., Berkoff T. A. and Morey W. W. "Multiplexed Fiber Bragg Grating Strain-Sensor System with a Fiber Fabry–Perot Wavelength Filter," Optics Letters, vol. 18, pp. 1370, 1993.

24. Hirose S., Yoneda K. "Development of 6-Axial Optical Force Sensor and Signal Decoding Method Under Simultaneous Multi-Load Conditions," Proc. SICE'89, 1989, pp. 735–736.

25. Peirs, J, Clijnen J, Reynaerts D, Brussel HV,et al. "A Micro Optical Force Sensor for Force Feedback During Minimally Invasive Robotic Surgery," Sensors & Actuators A Physical, 115, pp. 447–455, 2004.

26. Silverbrook, K., Lapstun P. "Optical force sensor," ed: US, 2010.

27. Kane, J., "Fiber Optics and Strain Interferometry," IEEE Transactions on Geoscience Electronics, 4, pp. 1–11, 2007.

28. Singh, J., Potgieter J., Xu W. "Fibre Optic Force Sensor for Flexible Bevel Tip Needles in Minimally Invasive Surgeries," International Journal of Biomechatronics and Biomedical Robotics, vol. 2, pp. 135–140, 2013.

29. He X., Handa J, Gehlbach P, Taylor R, Iordachita I. "A Submillimetric 3-DOF Force Sensing Instrument With Integrated Fiber Bragg Grating for Retinal Microsurgery," IEEE Transactions on Biomedical Engineering, vol. 61, pp. 522–534, 2014.

30. Bajo A, Simaan N, "Hybrid motion/force Control of Multi-Backbone continuum Robots," International Journal of Robotics Research, vol. 35, 2015.

31. Begej, S "Planar and Finger-Shaped Optical Tactile Sensors for Robotic Applications," IEEE Journal of Robotics & Automation, vol. 4, pp. 472–484, 1988.

32. Bakalidis GN Glavas E, Voglis NG, Tsalides P. "A Low-Cost Fiber Optic Force Sensor," IEEE Transactions on Instrumentation & Measurement, vol. 45, pp. 328–331, 1996.

33. Arminger B. R., Oppermann K., Zagar B. G. "Measurement of Highly Dynamic Forces Using an Elasto-Optical Force Sensor," Elektrotechnik und Informationstechnik, 127, pp.274, 2010.

34. Song, H., Kim K, Lee J. "Development of Optical Fiber Bragg Grating Force-Reflection Sensor System of Medical Application for Safe Minimally Invasive Robotic Surgery," Review of Scientific Instruments, 82, pp. 074301, 2011.

35. Tosi D. "Review and Analysis of Peak Tracking Techniques for Fiber Bragg Grating Sensors," Sensors, 17, pp. 2368, 2017.

36. Allwood, G., G. Wild, S. Hinckley, "Fiber Bragg Grating Sensors for Mainstream Industrial Processes," Electronics, 6, pp. 92, 2017.

37. Kim C and Lee C. H. "Development of a 6-DoF FBG Force–Moment Sensor for a Haptic Interface with Minimally Invasive Robotic Surgery," Journal of Mechanical Science & Technology, 30, pp. 3705–3712, 2016.

38. Marques C, Caucheteur C, Saezrodriguez D, Webb DJ, et al. "Polarization Effects in Polymer FBGs: Study and Use for Transverse Force Sensing," Optics Express, 23, pp. 4581–4590, 2015.
39. Raibert MH, and Craig JJ "Hybrid position/force Control of Manipulators," Asme J of Dynamic Systems Measurement & Control, vol. 102, pp. 126–133, 1981.
40. Guo Y, Kong J, Liu H, Xiong H, Li G, Qin L. "A Three-Axis Force Fingertip Sensor Based on Fiber Bragg Grating," Sensors & Actuators A Physical, vol. 249, pp. 141–148, 2016.
41. Melchiorri C, L. Moriello, G. Palli, U. Scarcia, "A new force/torque sensor for robotic applications based on optoelectronic components," In Robotics and Automation (ICRA), 2014 IEEE International Conference on pp. 6408–6413, 2014.
42. Palli G, Moriello L, Scarcia U, Melchiorri C. "Development of an Optoelectronic 6-Axis force/torque Sensor for Robotic Applications," Sensors & Actuators A Physical, vol. 220, pp. 333–346, 2014.
43. Su H, Zervas M, Cole G. A., Furlong C, Fischer G. S. "Real-Time MRI-Guided Needle Placement Robot With Integrated Fiber Optic Force Sensing," in IEEE International Conference on Robotics and Automation, 2011, pp. 1583–1588.
44. Tan U. X., Yang B, Gullapalli R, Desai J. P. "Tri-Axial MRI Compatible Fiber-Optic Force Sensor," IEEE Transactions on Robotics, 27, pp. 65, 2011.
45. W. Zhang, E. Li, J. Xi, J. Chicharo, and X. Dong, et al. "Novel Temperature-Independent FBG-Type Force Sensor." Measurement Science & Technology 16.5 (2005): 1600.2005.
46. Saccomandi P, Schena E, Oddo CM, Zollo L, et al. "Microfabricated Tactile Sensors for Biomedical Applications: A Review," Biosensors, 4, pp. 422–448, 2014.
47. Trejos AL, Patel RV, Naish MD "Force Sensing and Its Application in Minimally Invasive Surgery and Therapy: A Survey," ARCHIVE Proceedings of the Institution of Mechanical Engineers Part C Journal of Mechanical Engineering Science 1989-1996 (vols 203-210), 224, pp. 1435–1454, 2010.
48. Puangmali P., Liu H, Seneviratne L. D., Dasgupta P, Althoefer K, "Miniature 3-Axis Distal Force Sensor for Minimally Invasive Surgical Palpation," IEEE/ASME Transactions on Mechatronics, vol. 17, pp. 646–656, 2012.
49. Haslinger R., Leyendecker P, Seibold U. "A fiberoptic force-torque-sensor for minimally invasive robotic surgery," in IEEE International Conference on Robotics and Automation, 2013, pp. 4390–4395.
50. Polygerinos P., Ataollahi A, Schaeffter T, Razavi R, Seneviratne L. D., Althoefer K. "MRI-Compatible Intensity-Modulated Force Sensor for Cardiac Catheterization Procedures," IEEE Transactions on Biomedical Engineering, vol. 58, pp. 721–726, 2011.
51. Polygerinos P, Seneviratne L. D., Razavi R, Schaeffter T, Althoefer K. "Triaxial Catheter-Tip Force Sensor for MRI-Guided Cardiac Procedures," IEEE/ASME Transactions on Mechatronics, vol. 18, pp. 386–396, 2012.

52. Gonenc B, Chamani A, Handa J, Gehlbach P, et al. "3-DOF FORCE-SENSING MOTORIZED MICRO-FORCEPS FOR ROBOT-ASSISTED VITREORETINAL SURGERY," IEEE Sensors Journal, vol. PP, pp. 1–1, 2017.

53. Xu, K. J., Li C. "Dynamic Decoupling and Compensating Methods of Multi-Axis Force Sensors," IEEE Transactions on Instrumentation & Measurement, vol. 49, pp. 935–941, 2000.

54. Li, T., Shi C, Tan Y, Zhou Z, "Fiber Bragg Grating Sensing-Based Online Torque Detection on Coupled Bending and Torsional Vibration of Rotating Shaft," IEEE Sensors Journal, 17, pp. 1999–2007, 2017.

55. Roesthuis R. J., Kemp M., Dobbelsteen J. J. V. D., Misra S, "Three-Dimensional Needle Shape Reconstruction Using an Array of Fiber Bragg Grating Sensors," IEEE/ASME Transactions on Mechatronics, 19, pp. 1115–1126, 2014.

56. Park YL, Elayaperumal S, Ryu S, Daniel B, et al., "MRI-compatible Haptics: Strain sensing for real-time estimation of three dimensional needle deflection in MRI environments," International Society for Magnetic Resonance in Medicine (ISMRM), 17th Scientific Meeting and Exhibition, 2008.

57. Malla, R. B., Sen A and Garrick N. W., "A Special Fiber Optic Sensor for Measuring Wheel Loads of Vehicles on Highways," Sensors, 8, pp. 2551, 2008.

58. Park Y. L., Ryu S. C., Black R. J, Chau K. K, et al. "Exoskeletal Force-Sensing End-Effectors With Embedded Optical Fiber-Bragg-Grating Sensors," IEEE Transactions on Robotics, 25, pp. 1319–1331, 2009.

CHAPTER 4

Decoupling Algorithms

4.1 THE PRINCIPLE OF THE DECOUPLING

Multi-component Force/Torque (F/T) sensing systems have infiltrated a wide variety of automation products since the 1980s. Because the sensing systems adopt a monolithic elastic element to monitor multiple components of force terms along x-, y-, and z-axis (Fx, Fy, and Fz) and the torque terms about x-, y-, and z-axis (Mx, My, and Mz) simultaneously, coupling error occurs among the components(especially between component Fx and component My, component Fy and component Mx respectively).

Therefore, careful calibration and accurate decoupling are crucial aspects of multi-component F/T sensing systems. Specifically, the calibration experiment establishes the relationship between the applied load and the corresponding output voltages. While the decoupling process eliminates the unexpected output due to the coupling.

Several decoupling methods have been proposed and implemented for multi-component F/T sensing systems. The coupling methods mainly focus on two aspects: (1) designing monolithic elastic elements with novel structures [1] and manufacturing technologies [2] which aims to directly eliminate the root of coupling. For example, Song *et al.* [3] designed and fabricated a novel self-decoupled four-component wrist F/T sensor, which can be called as direct output force sensor. Wu *et al.* [4] analyzed the decoupling principle of a sliding six-component F/T sensor and proposed a robust design method of elastic body size optimization. Many mechanically decoupled force sensing systems such as systems based on the Steward platform [5, 6] were proposed. (2) proposing effective decoupling

TABLE 4.1   Researches about the decoupling of multi-component force sensor in recent years

Researcher	Decoupling Method	Experiment Object	Decoupling Effect
Krouglicof, Nicholas [7]	The decoupling matrix determined through finite element analysis	A six DOF, Stewart platform type force sensor	The overall accuracy was less than 0.4% of full scale over the entire operating range
Jianhe Lei [8]	BP neural network	An underwater robot wrist force sensor, six DOF	The I-type error ≤ 0.06%, and the II-type error ≤ 0.02%
Guo Jierong [9]	Least square support vector machine (LS-SVM)	A photoelectric displacement sensor	Less than 1.5%
Xiao Wenbin [10]	RBF neural network	A six-dimension force sensor	overall error is less than 1% full scale
Junqing Ma [11]	Based on coupling error model and ε-SVR	3-axis force sensor	All interference errors are reduced to less than 1.6%.
Ma Yingkun [12]	Based on support vector machines	A six-component force/torque transducer	errors are less than 0.1%

algorithms, i.e., finding the relationship mapping between the system output signal and the applied force/torque value. Compared with the former approach, the latter one is more convenient to be implemented, which can not only reduce the manufacturing requirement of the system but also be able to obtain the desired decoupling effect. Several related research about the decoupling algorithms of the multi-component force sensing systems in recent years are shown in Table 4.1.

Generally, decoupling algorithms can be classified into linear or nonlinear decoupling categories. Comparing with the latter one, linear decoupling methods would usually result in poorer precision because the actually existing relationship between the applied load and output of the multi-component force sensing systems is nonlinear. Thus, nonlinear decoupling methods have attracted particular attention. Commonly, intelligent algorithms such as neural networks are well used to approximate the input and output of the nonlinear relationship. Unfortunately, the decoupling speed is too slow to satisfy the requirement of a multi-component force sensing system's dynamic measurement. In recent years, the Extreme Learning Machine algorithm [13], proposed by Professor Huang from Nanyang Te chnological University, has presented fast speed and is able to achieve satisfying generalization performance compared with the traditional neural network, which induces the idea of nonlinear decoupling base on the ELM (Extreme Learning Machine).

This chapter will carry out calibration experiments and then implementing different linear and nonlinear decoupling methods for multi-component F/T sensing system with a double-E-membrane-type elastic element, respectively. Comparisons among decoupling methods on the aspects of measurement accuracy, coupling rate, and the decoupling speed will be provided.

4.2 TRADITIONAL LINEAR DECOUPLING

For static linear decoupling, the relationship between the applied load and the corresponding output of the multi-component force sensing system is assumed to be linear, which can be described as follows

$$F_{6 \times n} = C_{6 \times 6} \cdot U_{6 \times n} \tag{1}$$

where $F_{6 \times n}$ is the input matrix being comprised of n applied force/torque vector $[F_x, F_y, F_z, M_x, M_y, M_z]^T$, $U_{6 \times n}$ is the corresponding output matrix combined by n output voltage column vector $[U_{Fx}, U_{Fy}, U_{Fz}, U_{Mx}, U_{My}, U_{Mz}]^T$, and $C_{6 \times 6}$ is the calibration matrix of a multi-component force sensing system, which can convert the sensor output voltage readings to the force/torque value.

The static linear decoupling methods refer to solve the calibration matrix C, basically containing the following solving methods: [8].

4.2.1 Direct Inverse Decoupling ($n=6$)

Six sets of linearly independent force/torque column vectors were selected, which would then be used for calibration experiments. According to Eq. (1), the calibrating data samples can form 6 equations which are linearly independent. Finally, a sole calibration matrix C can be solved based on Cramer's Rule. The calculation formula can be rewritten as

$$C = FU^{-1} \tag{2}$$

This direct inverse decoupling method is convenient to calculate, but the method has a less decoupling precision due to the random errors and nonlinear relationship between the input and output in the calibration experiments.

4.2.2 Least Square Decoupling ($n>6$)

As large random errors exist in the calibration experiments, the calibration frequency generally needs to be increased. In this case, the acquired equations according to Eq. (1) would be more than 6. As a result, the above equations will have no solution, but its least squares-solution can be viewed as the calibration matrix. The calculation formula can be rewritten as

$$C = FU^{T}(UU^{T})^{-1} \qquad (3)$$

The linear decoupling based on the least square method actually makes a significance test for the calibration data covering the entire range of the sensor and then eliminates the deviation data, aiming to linearly approximate the sample data with the minimum error. Although it has partly eliminated the random error and overcome some nonlinear relationship to a certain extent, this method essentially belongs to the linear decoupling.

Generally, static decoupling methods are hard to obtain a satisfactory effect. This is mainly attributable to the nonlinear relationship between the output voltage and the applied force/torque of the multi-component force-sensing systems. In addition to that, the coupling between output channels is not entirely linear. So nonlinear decoupling methods take a more reasonable assumption that the relationship between the output and the input is nonlinear, trying to seek out a nonlinear mathematic model for the approximation of the nonlinear relationship, which can achieve an obviously better decoupling effect. Nowadays, the nonlinear approximation method mainly depends on the Neural Network or the Support Vector Regression. However, they are still hard to satisfy the dynamic requirement of the force-sensing systems due to their long time consumption during the network training processes. Therefore, the decoupling method based on the Extreme Learning Machine algorithm, which has a faster learning speed and stronger generalization ability, can become a new choice for nonlinear decoupling method for multi-component force sensing systems.

4.3 DECOUPLING BASED ON BP NEURAL NETWORK (BPNN)

The Back-Propagation Network is named after its main feature of the propagation algorithm. On one hand, the input information is propagated backward from the input layer passing hidden layers step by step. On the other hand, error signal is transferred from the output layer going through hidden layers. Meanwhile, the connection weights are gradually revised with criterion strategy, such as the Least Mean Square Algorithm. As the network learning process goes on, weighs are constantly adjusted and the final errors between the network output and the target output will get smaller and smaller. The training or iteration process of the network will not ease until the satisfied accuracy is reached. The last obtained weight coefficient will determine the BP neural network model which establishes the nonlinear bridge between the input and output of the multi-component force sensor, i.e., the nonlinear static decoupling.

The decoupling model based on BP neural network takes the column vector comprised of 6 channels of an output voltage of the sensor ($U_{6\times n}$) as the BP network's input, and the 6 corresponding F/T loaded on the sensor ($F_{6\times n}$) as the output. BP neural network generally takes the Tansig function as the transfer function of the hidden layer, while the output layer using Purelin linear function. And the Tansig function is smooth, differentiable, making its approximating nonlinear relationship in satisfactory fault tolerance. In addition, the hidden layer of BP network is set as a single layer, and its neuron number is determined by experiments.

Before the training process of the BP network, the initial weights are commonly randomly chosen, which results the non-repeatability of the network. Too large or too small initial weights will bring negative effects on network performance. Some intelligent optimization algorithms proposed by some researchers can be used to optimize the initial weights and threshold valve, making a better approximating accuracy with the BP decoupling method. But the complex optimization algorithm will pay the cost of a long decoupling time, which is bad for the dynamic response of the system.

4.4 DECOUPLING BASED ON SUPPORT VECTOR REGRESSION (SVR)

Support Vector Machine (SVM), developed by Vapnik for the applications of pattern recognition and nonlinear regression, can be divided as Support Vector Classification (SVC) and Support Vector Regression (SVR). As the decoupling of the multi-component force sensor is actually to explore the nonlinear relationship between the multi-channel output voltage signal and the applied multi-component load, the SVR algorithm can be adopted to decoupling of multi-component F/T sensing system.

The decoupling method based on SVR can be summarized as follows.

Given a set of training data set, $S=\{(x_i, y_i), x_i \in R^d, y_i \in R, i=1, 2, ..., n\}$, where x_i is a d-component input vector, y_i is the target output, and n is the number of data sample. Firstly, a nonlinear mapping is established as $x \rightarrow \varphi(x)$, which can change the low-dimensional input space into a high-dimensional space, and the nonlinear regression problem into a linear regression problem. Therefore, a regression function $f(x)$ can be obtained, through which the deviation of every sample's output value should be less than ε compared with the target output. In addition, the regression function needs to be as smooth as possible.

Assuming that the regression function can be written as

$$f(x) = w \cdot \varphi(x) + b \tag{4}$$

where w is the weight vector, and b is the threshold value. According to the basic idea of SVR, it can be described as a convex optimization problem with mathematical language:

$$\min \quad \frac{1}{2}\|w\|^2 + C\sum_{i=1}^{n}(\xi_i + \xi_i^*)$$

$$\text{s.t.} \quad \begin{cases} y_i - (w \cdot \varphi(x_i) + b) \leq \varepsilon + \xi_i \\ (w \cdot \varphi(x_i) + b) - y_i \leq \varepsilon + \xi_i^* \quad i=1,2,\cdots,n \\ \xi_i, \xi_i^* \geq 0 \end{cases} \tag{5}$$

where $\varepsilon \geq 0$ is the deviation error between the SVR predicting value and target value. The slack variables ξ_i and ξ_i^*, being introduced at the consideration of the possible big error existing in the training sample, reflect the error size of the corresponding data sample. The greater the slack variables are, the further distance the very sample deviates the overall trend. C is the penalty coefficient that determines the trades-off between the empirical risk and the regularization term.

In order to solve the Eq.5, the Lagrange multiplier (α_i and α_i^*) is introduced, and the above convex optimization problem can be transformed into its dual problem as follows:

$$\max_{\alpha_i,\alpha_i^*} \quad -\frac{1}{2}\sum_{i,j=1}^{n}(\alpha_i-\alpha_i^*)(\alpha_j-\alpha_j^*)[\varphi(x_i)\cdot\varphi(x_j)]-\varepsilon$$

$$\sum_{i=1}^{n}(\alpha_i+\alpha_i^*)+\sum_{i=1}^{n}y_i(\alpha_i-\alpha_i^*)$$

$$\text{s.t.} \quad \begin{cases} \sum_{i=1}^{n}(\alpha_i-\alpha_i^*)=0 & i=1,2,\cdots,n \\[2ex] \alpha_i,\alpha_i^* \in [0,C] \end{cases}$$

(6)

It can be seen that the above equation can be solved only if the inner product of $\varphi(x_i)$ and $\varphi(x_j)$ is calculated, without the necessity of finding the detail mapping form about $x\rightarrow\varphi(x)$. Therefore, a kernel function as follows needs to be determined.

$$K(x_i,x_j)=\varphi(x_i)\cdot\varphi(x_j) \tag{7}$$

Among the kernel function, the input parameters x_i and x_j are two low-dimensional space vectors, while the output value is the inner product of the two mapped high-dimensional space vectors. With the introduction of kernel function, a nonlinear regression problem in low-dimensional space can be changed into a linear regression problem in high-dimensional space. At present, the Radial Basis Function (RBF) is commonly used as the kernel function. This is because the feature space can be transformed into an infinite-dimensional space in which any limited data samples would be linear. The RBF function can be expressed as

$$K(x,x_i)=\exp\left(-\gamma\|x-x_i\|^2\right), \quad \lambda>0 \tag{8}$$

where γ is a parameter of the RBF kernel function, which needs to be determined by user.

Solving the Eq. 6, the Lagrange multiplier α_i, α_i^* can be obtained. Among them, the samples corresponding with $\alpha_i - \alpha_i^* \neq 0$ refer to Support Vectors. Then the weight vector w can be yielded as

$$w = \sum_{i=1}^{n} (\alpha_i - \alpha_i^*) \varphi(x_i) = \sum_{i=1}^{nSV} (\alpha_i - \alpha_i^*) \varphi(x_i) \tag{9}$$

where the nSV is the number of the Support Vectors. And the threshold value b can be solved out according to the Karush-Kuhn-Tucker (KKT) conditions.

$$b = y_i - w \cdot \varphi(x_i) - \varepsilon \qquad\qquad 0 < \alpha_i < C, \alpha_i^* = 0 \tag{10}$$

or

$$b = y_i - w \cdot \varphi(x_i) + \varepsilon \qquad\qquad 0 < \alpha_i^* < C, \alpha_i = 0 \tag{11}$$

Generally, the b is finally determined as the average value of all the b values which are calculated with every support vector.

$$b = \frac{1}{nSV} \left\{ \sum_{0 < \alpha_i < C} (y_i - w \cdot \varphi(x_i) - \varepsilon) + \sum_{0 < \alpha_i^* < C} (y_i - w \cdot \varphi(x_i) + \varepsilon) \right\}$$

$$= \frac{1}{nSV} \left\{ \sum_{0 < \alpha_i < C} (y_i - \sum_{j=1}^{nSV} (\alpha_j - \alpha_j^*) K(x_j, x_i) - \varepsilon) + \right.$$

$$\left. \sum_{0 < \alpha_i^* < C} (y_i - \sum_{j=1}^{nSV} (\alpha_j - \alpha_j^*) K(x_j, x_i) + \varepsilon) \right\} \tag{12}$$

The regression function can be eventually obtained as follows:

$$f(x) = w \cdot \varphi(x) + b = \sum_{i=1}^{n} (\alpha_i - \alpha_i^*) K(x_i, x) + b \tag{13}$$

As it can be seen from the Eq. 13, the regression function $f(x)$ can be acquired without the necessity of particularly calculating the weight vector w and the nonlinear mapping function $\varphi(x)$, but only the need the Lagrange multiplier (α_i, α_i^*), the kernel function $K(x_i, x)$, and the threshold value b. In addition, the complexity of the SVR function $f(x)$ has nothing to do with the dimensions of the input space, but only depends on the number of support vectors.

Through the above analysis, the nonlinear decoupling based on SVR algorithm can be summarized as follows:

Step 1: Get a training data sample set S;

Step 2: Select an appropriate precision parameter ε as well as the type of kernel function $K(x_i, y_i)$;

Step 3: Solve the optimization problem in formula (6) and yield $\alpha = (\alpha_1, \alpha_1^*, \cdots, \alpha_n, \alpha_n^*)$;

Step 4: Calculate the threshold value b;

Step 5: Construct a nonlinear SVR hyperplane $f(x)$.

In short, the SVR is similar in form to a neural network. The input of SVR network can be a multi-dimensional vector while the output is a single-dimensional real value linearly combined with some intermediate nodes, each of which corresponds to a support vector. And the weight coefficient is the Lagrange multiplier $(\alpha_i-\alpha_i^*)$. For the decoupling based on SVR, the six-dimensional output voltage vector is treated as the input of SVR model, while every single applied force or torque value is respectively taken as the

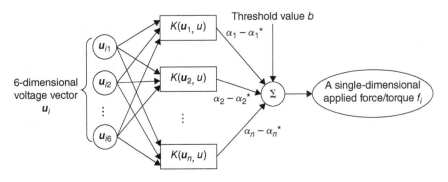

FIGURE 4.1   Schematic of decoupling based on SVR.

output. Therefore, six independent SVR models are constructed in turn and each one is used to predict the applied force/torque in every different direction. After the training, the target value (6-dimensional force/torque information) can be obtained with the six different SVR models.

4.5 DECOUPLING BASED ON EXTREME LEARNING MACHINE

Experiments show that, compared with the traditional linear decoupling, the decoupling precision is significantly improved as a result of the non-linear approximation capability of BP and SVR algorithms. However, BP neural network adopts the gradient descent method which needs a large number of iterations to revise the weights and bias value, so the training process would consume a long time. For the decoupling method based on the SVR, likewise, a lot of time is occupied in the procedure when the cross-validation method or a genetic algorithm is used for the selection of the primary parameters. A multi-component force sensor, however, needs a fast dynamic response when measuring besides meeting the static precision requirement. Recently, researches showed that the learning speed of ELM is further faster, and so is its generalization performance. Therefore, the decoupling method of a multi-component force sensor based on ELM algorithm is proposed.

The ELM algorithm actually refers to a special kind of single hidden layer neural network learning algorithm. It, different from the traditional BP algorithm, randomly generates the connection weights between the input layer and hidden layer as well as the bias of hidden layer neurons, all of which need no adjustment in the training process. After the setting of the number of hidden layer neurons, the optimal solution can then be acquired. Compared with the traditional training method, the significant advantage of the ELM algorithm is the extreme fast learning speed and its outstanding generalization performance. The principle of the ELM algorithm can be briefly described as follows.

The network structure of the decoupling model based on the ELM consists of an input layer, hidden layer, and output layer, in which the neurons are fully connected. The same as the BP and SVR decoupling, the ELM decoupling also adopts the reverse modeling, taking the 6 channels of output voltage as the model's input and the applied force/torque as the model's output. The number of the hidden layer neurons L needs to be determined in experiments.

The training procedure of the ELM decoupling can be concluded as follows:

1. Randomly generate the input weights ($w_{L\times6}$) and bias ($b_{L\times1}$) of the hidden layer neurons.

2. Calculate the hidden layer output matrix H.

3. Calculate the output weights of the hidden layer neurons with $\hat{\beta} = H^+ F$.

4.6 EXPERIMENTAL COMPARISON OF NONLINEAR DECOUPLING ALGORITHMS

Calibration experiments are to obtain a large amount of sample data with the applied force/torque information and the corresponding output voltage of the multi-component force sensor, aiming to provide the decoupling experiments with the training sample and testing sample.

4.6.1 Calibration Procedures

1. Keep the sensor well mounted on the calibration platform.

2. Divide the whole range scope of the sensor according to the loading weights, and obtain several equally spaced measuring points. Following the measuring points, load the weighs gradually increasing from zero to full scale value, and then decreasing return to zero. After that, reload the weights step-by-step increasing to negative full scale value, and finally decreasing to zero. In the loading process, the output voltage (value range is at ±5V) of each channel of the sensor can be recorded by the data acquisition software. The process stated above is called as a calibration circle.

3. Carry out the above calibration circle three times.

4.6.2 Analysis of the Calibration Experiment

Through the above calibration experiment, the relationship curve between the uniaxial loading force/torque and the corresponding output voltage can be obtained [14]. It is notable that, when the force/torque is applied along with the F_y, M_x or M_y direction, the coupling effects are produced on the M_z, F_y or F_x direction, respectively. All the coupling effects are mainly caused by the integrated elastic element structure, thus the decoupling process is essential.

4.6.3 Data Preparation of Decoupling Experiments

Among the sample data obtaining from the above calibration experiments, the force range value is less than 300, while some torque maximum value reached 10000, which exhibits a huge difference in magnitude. As this difference may result in a large prediction error of the network, the sample data needs to be normalized processed [15]. The data normalized processing is to convert all data into a number within the range of [−1, 1], aiming to cancel the differences of magnitude order between each dimension, which is also a preprocessing method generally doing before the neural network prediction. Some experiments also show that decoupling accuracy often tends to be better when the normalized procedure is conducted. Thus, before the decoupling experiments with every decoupling method in this paper, the sample data are all normalized processed. The data normalization methods are different from each other, one of which adopted in our experiments is the maximum-minimum method, whose function can be expressed by:

$$x_k = \frac{x_k - x_{min}}{x_{max} - x_{min}} \tag{23}$$

where the x_{min} is the minimum of the data series, and the x_{max} indicates the maximum.

The calibration experiments can generate six sets of data at the same loading condition. In order to guarantee the comprehensive nature of the sample data, this paper will take all the above calibration experiments data as the training sample. But for the testing sample which needs to be more accurate, the average value of the six groups of data under the same applied condition is calculated to eliminate the random error existing in the calibration experiments. In this way, the more accurate calibration sample data can be obtained and be used as a testing sample to justify the decoupling performance of different decoupling methods.

4.6.4 Decoupling Experiments

As shown in Fig 4.2, the linear decoupling and nonlinear decoupling experiments are both carried out. In order to compare the performance of different decoupling methods, the same testing sample data are adopted.

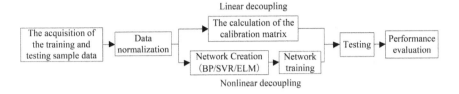

FIGURE 4.2 Decoupling procedure diagram of a multi-component F/T sensor.

And then the performance evaluation will be obtained from the aspects of measurement accuracy, static coupling rate, and the decoupling speed.

4.6.4.1 Linear Decoupling

Direct decoupling: Select six groups of sample data with the maximum range in each direction, and calculate the calibration matrix according to the formula (2). Then the calibration matrix is

$$
C_1 = \begin{bmatrix}
10.0134 & 2.9052 & -0.5232 & 3.3163 & -9.7937 & 0.7391 \\
2.0046 & 9.3709 & -0.4462 & 9.1660 & -2.0143 & 0.3859 \\
-0.1895 & -0.1696 & 1.0406 & -0.1918 & 0.2217 & -0.0546 \\
0.1056 & -0.0926 & -0.0656 & 0.9842 & -0.1107 & 0.0193 \\
0.0951 & 0.0606 & 0.0051 & 0.0818 & 0.9248 & 0.0066 \\
0.5542 & -1.0737 & -0.0788 & -0.3742 & -0.5408 & 1.6132
\end{bmatrix}
$$

Whose 2-norm is 17.28, and the mean square error (MSE) reaches to 1.26×10^5 when the C_1 is used to test the testing sample.

Least square decoupling: Take the whole training sample as the sample data, and calculate the least-square solution of the calibration matrix following formula (3). In this way, the calibration matrix can be calculated out as follows:

$$
C_2 = \begin{bmatrix}
6.5622 & 0.0215 & 0.0227 & -0.0446 & -6.5645 & 0.0229 \\
-0.1173 & 6.4783 & 0.0366 & 6.4507 & -0.0558 & -0.0531 \\
0.0473 & -0.0925 & 1.1133 & -0.0683 & 0.0704 & -0.0072 \\
-0.0200 & -0.1826 & -0.0004 & 0.7980 & 0.0052 & -0.0045 \\
0.1029 & 0.0064 & -0.0026 & 0.0149 & 0.8787 & -0.0035 \\
0.0472 & -1.2956 & 0.0057 & -0.7931 & -0.0747 & 1.0123
\end{bmatrix}
$$

Whose 2-norm is 9.33, and the MSE of the testing results is reduced to 3.21×10^3. From the 2-norm and MSE, it can be judged that the least square decoupling method has a better generalization performance than the direct decoupling.

4.6.4.2 Nonlinear Decoupling

Before the BP, SVR or ELM network is established, some parameters must need to be given first. Although the given parameters mostly exert more or less influence on the network training effects, there are no empirical methods to directly determine them. Thus, to improve the network generalization ability as much as possible, the traversal search way is adopted in this paper and the genetic algorithm, ant colony algorithm, particle swarm optimization algorithm as well as other intelligent optimization algorithms are considered as the feasible methods to determine the above parameters in our future research.

As for both the BP and ELM decoupling, the number of hidden layer neurons has a great impact on the prediction precision. If they were too few, the complex mapping relationship would be hard to build and the network's prediction error larger. On the contrary, if too many, the training time would increase and the over-fit phenomena may show up. In this paper, the number of the BP network's hidden layer neurons is initially chosen as 6~15 and similarly the ELM network as 20~50. Among the above neuron numbers, what we need is the one which can result in a less MSE and a short decoupling time. In this way, the hidden layer neuron numbers can be reliably determined in some sense.

As shown in Table 4.2, along with the increase of the hidden layer neurons number, the MSE of the output value will be gradually smaller while the decoupling speed will be slower due to the increasing calculation amount. For the trade-off of both the MSE and decoupling speed, the hidden layer neurons number of the BP decoupling network are finally determined as 13. Similarly, experiments also indicate that when the hidden layer neurons number of the ELM decoupling network is set at 38, the MSE of the predicted value has reached to a satisfied small value, and the corresponding decoupling time is also relatively short, as shown in Table 4.2.

As for the SVR, the penalty coefficient C and kernel function parameters γ are the main parameters, which need to be given before the training process. At present, a commonly adopted method is that, for the selected C and γ which are allowed within a certain range, the training set is regarded as the actual data set and the K-Cross-Validation (K-CV) method is adopted. In this way, the validation regression accuracy of the training set is obtained under different C and γ. Eventually, the very C

TABLE 4.2 Performance comparison of decoupling methods based on ELM and BP VS. Number of hidden layer neurons

BP Decoupling			ELM Decoupling		
Neuron Number	MSE ($\times 10^{-4}$V)	Decoupling Time (s)	Neuron Number	MSE ($\times 10^{-3}$V)	Decoupling Time (s)
6	1650	9	34	130	0.005
7	1013	4.8	35	121	0.002
8	507	3.9	36	147	0.002
9	263	7.7	37	168	0.006
10	305	4.1	38	120	0.002
11	324	1.8	39	142	0.003
12	291	0.9	40	150	0.005
13	256	0.6	41	147	0.005
14	267	7.8	42	140	0.005
15	251	6.1	43	146	0.005

and γ are determined when they make the validation regression accuracy of the training set the highest. The basic principle of the K-CV is simply described as follows: firstly divide the training set into K groups, and then respectively take each subset data as a test set while the other K-1 groups of the subset data as the training set, which can form the K different SVR models. And the average of the K models's prediction accuracy is considered as the SVR model's performance index. The optimal parameters obtained like this can effectively avoid the appearance of over-fitting or under-fitting to some extent, which are relatively more convincing. In the SVR decoupling process, it contains 6 SVR models, each of which needs an optimization parameter selection with K-CV method.

4.6.5 The Performance Evaluation of Decoupling

It is necessary to have an evaluation indicator to judge the different decoupling algorithms of multi-component force sensor. This paper takes the error rate stated in document [8] as a major evaluation indicator which is defined as

$$\text{Error Rate} = \left| \frac{F_{actual} - F_{predict}}{F_{full\text{-}scale}} \right| \tag{26}$$

where the F_{actual} is the force/torque value actually applied along each direction of the force sensor, $F_{predict}$ is the force/torque value obtained through decoupling process, $F_{full\text{-}scale}$ is the maximum force/torque in each direction of the sensor.

TABLE 4.3   The comparison of I-type and II-type relative error under the five different kinds of decoupling methods

Decoupling Methods	I-Type Relative Error (%)						II-Type Relative Error (%)
	F_x	F_y	F_z	M_x	M_y	M_z	Maximum Value
Direct reverse decoupling	**49.28**	31.72	0.66	2.90	2.62	7.60	48.80($M_z{\to}F_x$)
Least square decoupling	1.62	1.97	**4.16**	1.57	1.53	0.69	8.74($F_z{\to}F_y$)
BP decoupling	1.50	1.32	0.30	0.31	0.23	0.31	0.49($M_x{\to}F_y$)
SVR decoupling	1.02	0.94	0.44	0.71	**1.10**	0.62	0.89($M_x{\to}F_x$)
ELM decoupling	**0.92**	0.66	0.04	0.17	0.28	0.29	0.24($M_z{\to}F_y$)

According to the above definition of error rate, when the force is applied along the F_x direction, the error rate on all the six directions can be calculated through the formula (26). Among them, the error rate on F_x direction is called as I-type relative error, which reflects the deviation degree of the predict results compared with the actual applied value, also named the measuring precision. However, the error rates along the other five directions are named as II-type relative errors, which mirror the decoupling degree between each dimension, namely the static decoupling rate, which is the major factor resulting in the measuring error of a multi-component force sensor.

Based on the above analysis, I-type and II-type relative errors of the above five decoupling methods can be calculated out. To be benefit for the comparison, the average value of the I-type and II-type relative errors under the same applied direction are listed in Table 4.3.

The nonlinear decoupling algorithms (BP, SVR, and ELM) correspond an obviously much smaller error rates and deservedly possess better decoupling effects, compared with the linear decoupling methods (Direct reverse decoupling and the Least square decoupling). Further comparing the three kinds of the nonlinear decoupling, the error rates under the ELM algorithm are smaller than the BP and SVR.

To further compare the decoupling effects of the BP, SVR, and ELM algorithms, a detailed comparison figure about the error rates is given in Fig 4.3. As shown in it, the horizontal axis represents different applied points along a single direction, while the vertical axis indicates the error rates including the I-type and II-type error. The blue and green dotted lines respectively denote the BP and SVR decoupling, and the red solid line represents the ELM decoupling which is proposed in this paper. It can be seen from the picture that, the red solid line mostly appears below both

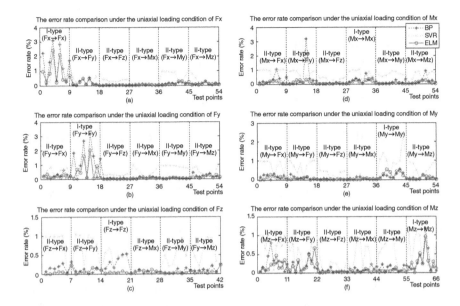

FIGURE 4.3 Relative error rates comparison of the BP, SVR and ELM decoupling algorithms under the uniaxial loading condition: (a) F_x; (b) F_y; (c) F_z; (d) M_x; (e) M_y; (f) M_z.

the blue and green dotted line, which means a smaller relative error rates and further a higher decoupling accuracy.

Besides the error rate, the decoupling speed is another important evaluation indicator of the decoupling algorithm, as a fast speed is obviously benefit for the satisfaction of the multi-component force sensor's dynamic measurement. So the comprehensive performance of the above nonlinear decoupling methods can also be compared both from the MSE and the running time of the decoupling program, as listed in Table 4.4. Seen from it, compared with the BP and SVR decoupling, the decoupling algorithm based on ELM has a much faster running speed and a relatively less mean square error of the decoupling, i.e., a better performance.

TABLE 4.4 The comparison of the decoupling performance among the BP, SVR and ELM decoupling methods

	Hidden Layer Neurons' Number	Decoupling Time	Mean Square Error
BP decoupling	13	0.8262s	183.7128
SVR decoupling	-	1.0138s	160.7400
ELM decoupling	35	0.0026s	134.9855

4.7 RELATED RESEARCH WORK

Decoupling of multi-component force sensing systems have attracted massive attentions of researchers recently. Generally, the majority of the existing multi-component force sensing systems are confronted by high coupled interference errors among their components due to their monolithic elastic element structures, inherent manufacturing errors, complicated sensing principles, and signal processing algorithms. According to the coupling model of a multi-component force sensing system, the output signals of multiple components will be simultaneously generated when the system is applied with a single component load due to the coupling error. Therefore, the effective calibration and decoupling process of a multi-component force sensing system is critical and challenging. Existing decoupling process of multi-component force sensing systems always focus on hardware, software decoupling methods and their combination [15].

Dr. G. A. Kebede [16], National Taiwan University of Science and Technology, proposed a decoupled six-component force sensing system based on novel strain gauge arrangement strategy and error reduction techniques. The calibration and decoupling processes were carried out based on the Least Squares Method, and a decoupling matrix was used to relate the applied load and the corresponding output force information. Six-component force and torque measurement up to 98.1 N and 13.322 Nm show a crosstalk error decrease to 4.78% F.S.

At the School of Mechanical Engineering, University of Jinan, China, an static decoupling approach for multi-component piezoelectric force sensing system was proposed based on Least Square (LS) for decoupling and Least Square Support Vector Machine Regression (LSSVR) for optimize nonlinear characteristics of the system output [17]. Experimental results indicated that the maximum nonlinear and coupling errors were reduced to 0.89% F.S. and 0.1% F.S., respectively, which shows that the proposed decoupling approach is important for force sensing accuracy improvement.

A novel six-component force sensing system based on a cross-beam structure with sliding and rotating mechanisms has been investigated in Ref. [18]. Thanks to the unique sliding and rotating structures, the structural interconnection of the proposed force sensing system was reduced and then the decoupling effect was correspondingly reduced. Mathematical models, their idealizations, and the mechanical decoupling principle were provided. Theoretical analysis and numerical analysis were conducted by Timoshenko beam theory and finite element analysis, respectively.

In comparison with existing force sensing systems, the proposed system demonstrated a decreased coupling error (the maximum coupling error was under 2.12%, which occurred in F_x component under SM_y).

Static decoupling of a piezoelectric multi-component force sensing system with four point structure was investigated by means of BP forward feedback neural network [19], [20]. The coupling model of the piezoelectric multi-component force sensing system was constructed via a BP neural network with three layers. A network model of the (back propagation) BP was constructed with an algorithm toolbox, and the number of hidden layer element nodes was determined based on experience. The maximum I-type and II-type errors were 0.5% and 2.93% F.S., which proves that the proposed static decoupling algorithm was effective.

Y. Song et al. [21] proposed decoupling research based on an Improved BP network for a three-dimensional flexible tactile sensor. The mapping relationship between the applied force and the corresponding resistance variation was decoupled by the proposed algorithm. The gradient descent with the momentum factor method was adopted to improve the performance of the BP network. Finite element analysis was used to analyze and simulate the proposed flexible tactile sensor. Experimental results indicated that the proposed decoupling method is effective and the maximum coupling error was 2.61% F.S.

A novel static calibration and decoupling approach for six-component force sensing system was proposed based on genetic algorithm (GA) [22]. In comparison with least-squares method (LSM), the proposed GA-based decoupling process can greatly improve the measurement accuracy of the sensing system. The maximum I-type and II-type errors were less than 0.4% and 2.7%, respectively, which prove the effectiveness of the proposed decoupling approach.

4.8 CONCLUSION

The training and testing sample data are obtained through calibration experiments for a six-component F/T sensor. Two kinds of traditional static linear decoupling methods (direct reverse decoupling, least square decoupling) as well as three static nonlinear decoupling algorithms (BP, SVR and ELM) were applied for the decoupling experiments. Meanwhile, the decoupling performance evaluation is respectively implemented. Experimental results indicate that the nonlinear decoupling methods outperform the traditional linear decoupling methods.

Furthermore, through the comparison of the three kinds of nonlinear decoupling methods, the ELM has a relatively better decoupling accuracy than the BP and SVR-based coupling methods. What's more, both the BP- and SVR-based decoupling methods are too slow to satisfy the dynamic measuring requirement of the multi-component force sensor due to the redundant iterations or the complex parameter optimization process. Therefore, the newly proposed decoupling method based on ELM for multi-component F/T sensor not only has a high decoupling precision, but also exhibits a fast decoupling speed which is a benefit to the improvement of the measuring accuracy and dynamic response.

However, as found in the experiments, the decoupling effects with ELM are satisfied under the uniaxial loading condition, but not be the same perfect when the combined loads are applied, which is partly due to the inadequate learning samples covering too narrow measuring range, and also partly because the proposed algorithm has a relatively large room for continuous improvement.

REFERENCES

1. Min-Kyung K, Soobum L, Jung-Hoon K. Shape Optimization of a Mechanically Decoupled Six-Axis force/torque Sensor [J]. Sensors and Actuators A (Physical). 2014, 209:41–51.
2. Xiaohong D, Weichao Y, Huanghuan S, Yong Y, Yunjian G, Jian S. "A Distributing and Decoupling Method of Microminiature Multi-Dimension Robot Finger Force Sensor [J]." Internationnal Conference on Robotics and Biomimetics. (2007): 1522.
3. Aiguo S, Juan W, Gang Q, Weiyi H. "A Novel Self-Decoupled Four Degree-of-Freedom Wrist force/torque Sensor [J]." Measurement 40.9-10 (2007): 883–91.
4. Bo W, Ping C. "Decoupling Analysis of a Sliding Structure Six-Axis force/torque Sensor [J]." Meas Sci Rev 13.4 (2013): 15–21.
5. T.A. Dwarakanath BD, Mruthyunjaya T.S. Design and Development of a Stewart Platform Based Force–Torque Sensor [J]. Mechatronics. 2001, 11:793–809.
6. Hou Y, Zeng D, Yao J, Kang K, Lu L, Zhao Y. "Optimal Design of a Hyperstatic Stewart Platform-Based force/torque Sensor With Genetic Algorithms [J]." Mechatronics 19.2 (2009): 199–204.
7. Krouglicof N, Alonso L.M., Keat W.D. Development of a Mechanically Coupled, Six Degree-of-Freedom Load Platform for Biomechanics and Sports Medicine [J]. 2004, - 5(-):- 4431,5.
8. Jianhe, L, Q LianKui, L Ming, S Quanjun, G Yunjian. - Application of Neural Network to Nonlinear Static Decoupling of Robot Wrist Force Sensor [J]. 2006,- 2(-):- 5285.

9. Guo Jie-rong HY-g, Liu Chang-qing. Nonlinear Correction of Photoelectric Displacement Sensor Based on Least Square Support Vector Machine [J]. J Cent South Univ Technol. 2011(18):1614–8.
10. Xiao, W-b, C Huo, W-c Dong. Research on Static Characteristics of Six-Dimension Force Sensor. 2012,-(-):- 581.
11. Junqing M, Aiguo S, Jing X. "A Robust Static Decoupling Algorithm for 3-Axis Force Sensors Based on Coupling Error Model and Epsilon-SVR [J]." Sensors 12.11 (2012): 14537–55.
12. Ma Y, Xie S, Zhang X, Luo Y. "Hybrid Calibration Method for Six-Component force/torque Transducers of Wind Tunnel Balance Based on Support Vector Machines [J]." Chinese Journal of Aeronautics 26.3 (2013): 554–62.
13. Huang G-B, Qin-Yu Zhu, and Siew Chee-Kheong. "Extreme Learning Machine: Theory and Applications [J]." Neurocomputing 70.1 (2006): 489–501.
14. Liang Q, Zhang Dan, Song Quanjun, Ge Yunjian. "Design and Fabrication of a Six-Dimensional Wrist force/torque Sensor Based on E-Type Membranes Compared to Cross Beams [J]." Measurement 43.10 (2010): 1702–19.
15. Liang Q, Wu W, Coppola G, Zhang D, et al. "Calibration and Decoupling of Multi-Axis Robotic Force/Moment Sensors." Robotics and Computer-Integrated Manufacturing 49 (2018): 301–308.
16. Kebede G.A, Ahmad A.R, Lee S.C, Lin C.Y. "Decoupled Six-Axis Force–Moment Sensor With a Novel Strain Gauge Arrangement and Error Reduction Techniques." Sensors 19.13 (2019): 3012.
17. Li Ying-jun, Wang Gui-cong, Yang Xue, Zhang Hui, et al. "Research on Static Decoupling Algorithm for Piezoelectric Six Axis force/torque Sensor Based on LSSVR Fusion Algorithm." Mechanical Systems and Signal Processing 110 (2018): 509–520.
18. Lin C.Y, Ahmad A.R, Kebede G.A and Lee S.C. "Novel Mechanically Fully Decoupled Six-Axis Force-Moment Sensor." Sensors 20.2 (2020): 395.
19. Li, Yingjun, Wang, Guicong, Han, Binbin, Yang, Xue, et al. "Research on Nonlinear Decoupling Method of Piezoelectric Six-Dimensional Force Sensor Based on BP Neural Network." MS&E 428.1 (2018): 012041.
20. Li, Y.J., Han, B.B., Wang, G.C., Huang, S., et al. "Decoupling Algorithms for Piezoelectric Six-Dimensional Force Sensor Based on RBF Neural Network." Optics and Precision Engineering 25.5 (2017): 1266–1271.
21. Song Yang, Wang Feilu, and Zhang Zhenya. "Decoupling Research of a Novel Three-Dimensional Force Flexible Tactile Sensor Based on an Improved BP Algorithm." Micromachines 9.5 (2018): 236.
22. Li Hui-Jun, Song Ai-Guo, Li Ang, Fu Li-Yue. "Static Calibration and Decoupling for Six-Axis Force/Torque Sensors Based on Genetic Algorithm." Journal of Nanoelectronics and Optoelectronics 14.3 (2019): 431–441.

Appendix

APPENDIX A DECOUPLING METHOD BASED ON BPNN

```
Training_D=xlsread('Training_D.xls','A2:L301');
Vol_train=Training_D(:,1:6)';
FT_train=Training_D(:,7:12)';
Testing_D=xlsread('Testing_D.xls',A2:L55');
Vol_test=Testing_D(:,1:6)';
FT_test=Testing_D(:,7:12)';
[P_train,input] = mapminmax(Vol_train);
P_test = mapminmax('apply',Vol_test,input);
[FT_train,output] = mapminmax(FT_train);
FT_test = mapminmax('apply',FT_test,output);
i=1;
for Node_N=6:15
tic
net=feedforwardnet(Node_N);
[net,tr]=train(net,P_train,FT_train);
Tn_sim=sim(net, P_test);
T_sim = mapminmax('reverse',Tn_sim,output);
BP_time=toc;
result=[FT_test' T_sim'];
error=FT_test'-T_sim';
E_mse=mse(error);
disp(['Node_N ','E_mse ','BP_time ','Epochs']);
performance=[Node_N,E_mse,BP_time,tr.num_epochs];
Node_N0(i)=Node_N;
E_mse0(i)=E_mse;
BP_time0(i)=BP_time;
Epochs0(i)=tr.num_epochs;
i=i+1;
disp(performance);
if Node_N==13
```

```
result=[F_test' F_sim'];
error_mse=mse(error)
norm_C2=norm(C2)
Liang=ones(54,1)*[200 200 300 8000 8000 10000];
ErrorRate=abs(error./Liang);
end
end
subplot(1,2,1),plot(Node_N0,E_mse0,'r-x');
subplot(1,2,2),plot(Node_N0,BP_time0,'b-*');
```

APPENDIX B DECOUPLING METHOD BASED ON SVR

```
Training_D=xlsread('Training_D.xls','A2:L301');
Vol_train=Training_D(:,1:6)';
FT_train=Training_D(:,7:12)';
Testing_D=xlsread('Testing_D.xls',A2:L55');
Vol_test=Testing_D(:,1:6)';
FT_test=Testing_D(:,7:12)';
[P_train,input] = mapminmax(Vol_train,1,2);
P_test = mapminmax('apply',Vol_test,input);
P_train=P_train';
P_test=P_test';
[FT_train,output] = mapminmax(FT_train,1,2);
FT_test = mapminmax('apply',FT_test,output);
FT_train=FT_train';
FT_test=FT_test';
tic
for i=1:6
[bestmse,bestc,bestg]=SVMcgForRegress
(FT_train(:,i),P_train,-4,4,-4,4,3,1,1,0.1);
disp;
str = sprintf ('Best Cross Validation MSE = %g Best c
= %g Best g = %g',bestmse,bestc,bestg);
disp(str);
cmd = ['-c ', num2str(bestc), '-g ', num2str(bestg),
'-s 3 -p 0.01'];
model = svmtrain(FT_train(:,i),P_train,cmd);
[Tn_predict(:,i),mse1] =
svmpredict(FT_test(:,i),P_test,model);
str = sprintf ('MSE = %g R =
%g%%',mse1(2),mse1(3)*100);
disp(str);
```

```
end
SVM_T=toc
T_predict= mapminmax('reverse',Tn_predict',output);
result = [FT_test' T_predict']
error=FT_test'-T_predict';
E_mse=mse(error)
SVM_T
Liang=ones(54,1)*[200 200 300 8000 8000 10000];
ErrorRate=abs(error./Liang);
```

Index